POLLUTED & DANGEROUS

America's Worst Abandoned Properties and What Can Be Done About Them

Justin B. Hollander

University of Vermont Press
Burlington, Vermont

Published by University Press of New England
Hanover and London

University of Vermont Press
Published by University Press of New England,
One Court Street, Lebanon, NH 03766
www.upne.com

Printed in the United States of America

5 4 3 2 1

Library of Congress Cataloging-in-Publication Data

Hollander, Justin B.
  Polluted & dangerous : America's worst abandoned properties and what can be done
about them / Justin B. Hollander.
     p.   cm.
  Includes bibliographical references and index.
  ISBN 978-1-58465-719-4 (cloth : alk. paper)
  1. Industrial archaeology—United States. 2. Factories—Environmental aspects—
United States. 3. Waste lands—United States. 4. Abandoned buildings—United
States. 5. Brownfields—United States. 6. Urban renewal—United States.
I. Angulo Aguado, Irene. II. Title. III. Title: Polluted and dangerous.
  T21.H64 2008
  333.77'1370973—dc22   2008038611

All illustrations are by the author unless otherwise indicated.

University Press of New England is a member of the Green Press Initiative.
The paper used in this book meets their minimum requirement for
recycled paper.

I dedicate this work to the memory of Donald A. Krueckeberg (1938–2006).

Contents

List of Figures   ix
List of Tables   xi
Acknowledgments   xiii
List of Acronyms   xv

  1. Introduction   1
  2. Research, Writing, and Thinking on HI-TOADS   13
  3. Local Officials and Their Attitudes toward HI-TOADS   30
  4. A National Perspective on What Cities Do about HI-TOADS   47
  5. Redevelopment Policy in a Municipal Coalition City:
       Trenton, New Jersey   78
  6. Slag Heaps, Steel Mills, and Sears: Pittsburgh, Pennsylvania   118
  7. First Whales, Now Brownfields: New Bedford, Massachusetts   149
  8. Planning for a Shrinking City: Youngstown, Ohio   176
  9. Race, Preservation, and Redevelopment: Richmond, Virginia   203
 10. Conclusion   232
 Epilogue   252

APPENDIXES

APPENDIX A. Detailed Methodology for Identifying Cities for
    Inclusion in the Study   257
APPENDIX B. Ranking of U.S. Cities on Likelihood of Hosting
    HI-TOADS Using Five Key Variables   269
APPENDIX C. Correlation Matrix for Five Key Variables   271
APPENDIX D. Results of Cluster Analysis Using a Setting of
    Five Clusters   273
APPENDIX E. U.S. Cities with Populations Greater than 100,000
    in 1970   275
APPENDIX F. Interview Instrument   277

Contents

APPENDIX G. Detailed List of HI-TOADS   279

APPENDIX H. Case-Study Methodology   289

Notes   293

Bibliography   299

Index   311

Figures

Figure 3.1. Identification of Cities Relatively Likely to Have
Neighborhoods with HI-TOADS   33

Figure 4.1. Combined Use of Policy Tools to Address HI-TOADS   61

Figure 5.1. Allan Mallach, former City Planning Director,
Trenton, New Jersey   84

Figure 5.2. HI-TOADS in Trenton, New Jersey   86

Figure 5.3. Magic Marker Site   90

Figure 5.4. Magic Marker Site   91

Figure 5.5. Crane Site   93

Figure 5.6. Roebling Complex   98

Figure 5.7. Roebling Complex   99

Figure 5.8. Assunpink Greenway   100

Figure 5.9. Assunpink Greenway   101

Figure 6.1. HI-TOADS in Pittsburgh, Pennsylvania   122

Figure 6.2. South Side Works   126

Figure 6.3. South Side Works   127

Figure 6.4. Nine Mile Run   130

Figure 6.5. Hazelwood LTV Site   132

Figure 6.6: Hazelwood LTV Site   133

Figure 6.7. Sears Site   136

Figure 6.8. Sears Site   137

Figure 7.1. HI-TOADS in New Bedford, Massachusetts   154

Figure 7.2. Elco Dress Factory   155

Figure 7.3. Elco Dress Factory   156

Figure 7.4. Morse Cutting Tools   161

Figure 7.5. Morse Cutting Tools   162

Figure 7.6. Pierce Mills   164

Figure 8.1. HI-TOADS in Youngstown, Ohio   180
Figure 8.2. Aeroquip Site   184
Figure 8.3. Aeroquip Site   185
Figure 8.4. 229 East Front Street   187
Figure 8.5. Youngstown Sheet and Tube / Republic Steel   188
Figure 8.6. Master Plan for the Reuse of the YST/RS Site   189
Figure 8.7. YST/RS: Steelworkers Bridge   190
Figure 8.8. Neighborhood 87   194
Figure 9.1. HI-TOADS in Richmond, Virginia   207
Figure 9.2. Fulton Gas Works   212
Figure 9.3. Fulton Gas Works   213
Figure 9.4. Detail of 2000–2020 Richmond Master Plan   215
Figure 9.5. Richmond Memorial Hospital   217
Figure 9.6. Tobacco Row   219
Figure 9.7. Tobacco Row   220
Figure 10.1. Case-Study Sites: Policy Applications 1995 to 2005   235
Figure 10.2. Case-Study Sites: Neighborhood Performance 1990 to 2000   239

Tables

Table 1.1. Typology of Unoccupied Urban Land Uses   5

Table 3.1. Comparison between Cities that Participated in the Study
and Those that Did Not   35

Table 3.2. Descriptive Statistics of Key Independent and Dependent
Variables from Interviews   38

Table 3.3. Active or Inactive HI-TOADS from 1995 to 2005   40

Table 4.1. Policies Adopted to Address HI-TOADS   48

Table 4.2. Sum of Policies Adopted for Each City   62

Table 4.3. Cross-Tab between Planning and Site Work   63

Table 4.4. Cross-Tab between Planning and Condemnation   64

Table 4.5. Non-Industrial Land Uses Explored, Envisioned, Planned,
or Acted Upon at Former or Current HI-TOADS   65

Table 4.6. Cross-Tab between Zoning and Success   69

Table 4.7. Cross-Tab between Planning and Success   70

Table 4.8. Cross-Tab between Planning/Zoning and Success   71

Table 5.1. Summary Information for HI-TOADS in Trenton   88

Table 5.2. Property Value and Demographic Data for HI-TOADS
and the City of Trenton, 1980 to 2000   94

Table 6.1. Summary Information for HI-TOADS in Pittsburgh   121

Table 6.2. Property Value and Demographic Data for HI-TOADS
and the City of Pittsburgh, 1980 to 2000   123

Table 7.1. Summary Information for HI-TOADS in New Bedford   153

Table 7.2. Property Value and Demographic Data for HI-TOADS
and the City of New Bedford, 1980 to 2000   157

Table 8.1. Summary Information for HI-TOADS in Youngstown   179

Table 8.2. Property Value and Demographic Data for HI-TOADS
and the City of Youngstown, 1980 to 2000   181

Table 9.1. Summary Information for HI-TOADS in Richmond   206
Table 9.2. Property Value and Demographic Data for HI-TOADS
    and the City of Richmond, 1980 to 2000   208
Table A.1. Socio-Economic Profile of 153 Cities   259
Table A.2. Descriptive Statistics of Variables in Analyses   261
Table A.3. Table of Runs   262
Table A.4. Factor Loadings for Run #2   263
Table A.5. Geographic Distribution of 153 Cities in Population and
    30 Cities in Sample   265
Table A.6. Comparison between 153 Cities in Population and
    30 Cities in Sample   266

Acknowledgments

This work owes much to my friend and mentor Michael Greenberg. I greatly appreciate his insights and support throughout my doctoral education and since. The research upon which this book is partially based was conducted for a doctoral dissertation. I want to acknowledge my dissertation committee members, for whom I am very grateful: Frank Popper, Kris Wernstedt, and the late Don Krueckeberg. While at Rutgers, I had the pleasure of getting to know other faculty members, and I particularly want to acknowledge Hank Mayer, Karen Lowrie, Julia Rubin, Radha Jagannathan, Clint Andrews, Robert Burchell, Bob Lake, Dona Schneider, and David Listokin for their friendship and the important roles they played in preparing me for a career in academia.

I want to acknowledge the efforts of the over one hundred people who volunteered their time to speak with me about HI-TOADS. Special thanks also go to my research assistant, Erin Heacock.

This research was made possible through the financial support of fellowships from the Lincoln Institute of Land Policy (XJH052206) and the Appraisers' Research Foundation. In addition, I want to acknowledge the financial support I received from the Dean of Arts and Sciences and the emotional support offered by the faculty and staff in the Department of Urban and Environmental Policy and Planning at Tufts University.

I want to express my sincere appreciation to Glenn Rotondo, John Kelly, William Costa, and John Marcic at the U.S. General Services Administration. Their flexibility and support from 2003 to 2006 was essential in the completion of this study.

But most of all, I want to thank my family for their love and encouragement, without which this book would not have been possible. I especially want to thank Joel and Eleanor, Ben and Amy, Josh, Aliza, Rachayl, Mom and Dad, and Pam and Rose!

Acronyms

BDA	Brownfields Development Area
BEST	Brownfields Environmental Solutions for Trenton
BOA	Brownfields Opportunity Area
CBO	community-based organization
CDC	Community Development Corporation
CERCLA	Comprehensive Environmental Response, Compensation, and Liability Act
DCA	Department of Community Affairs
DECD	Department of Economic and Community Development
DEP	Department of Environmental Protection
ECRA	Environmental Cleanup Responsibility Act
EPA	(U.S.) Environmental Protection Agency
EWS	Early Warning System
FEMA	Federal Emergency Management Agency
GSA	General Services Administration
HDSRF	Hazardous Discharge Site Remediation Fund
HI-TOADS	High-Impact Temporarily Obsolete Abandoned Derelict Sites
HRS	Hazardous Ranking System
HUD	(U.S. Department of) Housing and Urban Development
ISRA	Industrial Site Recovery Act
LDC	Local Development Corporation
LULU	Locally Unwanted Land Use
MRCO	Mahoning River Corridor of Opportunity
NGO	nongovernmental organization
NJDEP	New Jersey Department of Environmental Protection
NOAA	National Oceanic and Atmospheric Administration

PCBs	polychlorinated biphenyls
PDD	Planned Development District
RIDC	Regional Industrial Development Corporation
SHPO	State Historic Preservation Officer
SSLDC	South Side Local Development Corporation
TOADS	Temporarily Obsolete Abandoned Derelict Sites
URA	Urban Renewal Authority
WHALE	Waterfront Historic Area League

[1]

Introduction

Built around 1943 to support the war effort, the Delco Appliance Factory is located between Orchard and Whitney Avenue in the Dutchtown neighborhood of Rochester, New York (Marcotte 2004). While only one of thousands of similar manufacturing sites across the United States, the Delco Factory stands as a useful point of beginning to introduce this book.

As the forces of globalization closed in on the automotive industry, the Delco Company closed the factory and emptied the heroic building around 1995. From 1950 to 2000, Rochester lost nearly one-third of its population, going from a city of 322,000 to a city of just 220,000 (U.S. Census 2000). Arson, crime, and dumping of waste around the Delco Factory spilled over into the Dutchtown neighborhood. Rochester's planning director estimated that properties more than a quarter mile away from the Delco Factory have been depressed due to the site. In a blighting fashion, the site spreads its impacts far beyond its immediate environs. Beyond the dangers of crime and abandonment, concerns about what and when some kind of new use will materialize at the site act as a chilling effect on further investment.

Just over a mile away, officials in city hall go about the ugly business of managing a declining American city. Ever optimistic, city leaders advance sustainability, economic development, education, and public safety agendas to revive their despondent hometown (Duffy 2006). The city boasts on its website about successful redevelopment projects, but what about the really serious sites like the Delco Factory that are literally dragging down their neighborhoods? What does the Delco Factory mean to the sustainability of Rochester? Do local officials think that this kind of site represents a real problem? If so, what do they do to address these kinds of sites and how successful are they? With environmental risks and redevelopment costs high, what can cities do?

Background on Vacant and Abandoned Land

Vacant and abandoned urban land is a widespread phenomenon afflicting neighborhoods in nearly every major U.S. city (Pagano and Bowman 2000). In some cases, these properties have or are perceived to have environmental contamination and are classified as "brownfields" (Bartsch and Collaton 1997). A particularly serious type of brownfield was identified by Greenberg and his colleagues in a 1990 study. These "temporarily obsolete abandoned derelict sites," or TOADS, are more dangerous than a typical brownfield because social and economic impacts are believed to stretch beyond their site boundaries, far into their surrounding neighborhood. As with any brownfield property, these TOADS had been in productive use but were abandoned by their owners and now sit idle. They are "scattered, random, unused parcels of land of varying size and shape . . . some have abandoned structures; others are only empty lots" (Greenberg, Popper, and West 1990, 435). In interviews, researchers asked New Jersey local tax assessors to estimate the impact these TOADS had on neighborhood property values or property transactions. Using the criteria that a site was just a brownfield if it was idle and had real or perceived contamination, it was a TOAD if the site's abandoned status was likely causing impacts on property values or transactions off-site, and it was a high-impact TOAD (HI-TOAD) if those impacts were estimated to extend beyond one-quarter mile away from the site's boundaries, the study found that a third of all municipalities in the state have brownfields (Greenberg et al. 2000). It also showed that 10 percent of municipalities have TOADS, and 3 percent have HI-TOADS. These worst-of-the-worst brownfield sites, these HI-TOADS, will be the focus of this book. Strictly speaking, HI-TOADS are real properties, without legal occupants and protection, which are estimated by local officials to harm property values more than one-quarter mile away. When HI-TOADS no longer meet this definition, because an owner enacts protective measures at the property, a new use at the site brings occupation, or any other means, then I consider it to be converted successfully and no longer a HI-TOAD site.

Research cited above suggests that economically distressed cities are most likely to have HI-TOADS. While HI-TOADS could be found in rural or suburban areas, the focus of this book will be on urban areas. The Greenberg's 2000 study predicts that 17 of New Jersey's 566 municipalities

(3 percent) should be expected have neighborhoods with HI-TOADS and 186 should be expected to have neighborhoods with brownfields. Using that same ratio of brownfields to HI-TOADS and Simons' (1999) estimate of 500,000 brownfields in the United States, then there could well be thousands of HI-TOADS in the United States.

People living near HI-TOADS can suffer tremendously from the criminal, public health, and environmental risks the sites may pose. Moreover, such people tend to be disproportionately poor and non-white, groups who traditionally have had little power to fight the effects of blight (Greenberg et al. 2000; Young 2000). Such conditions make HI-TOADS environmental and social justice issues and a key barrier to creating sustainable communities.

Market conditions tend to play an important role in determining which HI-TOADS get redeveloped and which get passed over. The attractiveness of a site to capital investment may depend on how investors view its neighborhood. For that reason, the fortunes of HI-TOADS are tied very closely to their neighborhoods' economic vitality. Local intervention may attempt to stimulate neighborhood market strength to make a HI-TOADS redevelopment project profitable for investors.

The purpose of the book is to examine this market intervention and other ways that cities address these HI-TOADS in urban areas; in particular, what works and what does not. In doing so, this book will address the challenge of achieving sustainability in America's distressed cities.

Unoccupied Urban Land Use Types

While a unique concept, HI-TOADS are characterized by the worst elements of other types of urban unoccupied land uses: vacant lots, abandoned buildings, and brownfields. In this section, I describe the relationship of HI-TOADS to each of the other classifications. Returning to the definition for HI-TOADS above, I focus exclusively on properties that are not being occupied *legally*. An illegal occupation, for the purposes of this research, is the same as an unoccupied site. Before delving into unoccupied land uses, I first will contrast unoccupied land uses with occupied land uses that also generate neighborhood impacts.

Locally Unwanted Land Uses (LULUs) can come in many forms, from slaughterhouses to refineries to garbage dumps (Popper 1981). In many cities, for such uses the operating plant causes more problems to residents than the abandoned plant, but not always. And most importantly, no matter how noxious and unpleasant a LULU may be, operating sites almost always have some redeeming value to cities in terms of local tax revenues and jobs. They also are needed regionally, despite being unwanted locally. The unoccupied land uses below are different from LULUs in that they cause harm to surrounding areas, yet offer no positive benefits to nearby residents or area governments.

Pagano and Bowman (2000) conducted research recently on vacant lots, sending a survey to local officials with the intent to "gather information about vacant land and abandoned structures" (2). The definition they used was: "Vacant land includes not only publicly owned and privately owned unused or abandoned land or land that once had structures on it, but also the land that supports structures that have been abandoned, derelict, boarded up, partially destroyed, or razed" (2).

While their definition was useful for their purposes, abandoned buildings are separate from vacant lots for the purposes of this typology. Vacant lots are conceptually independent from abandoned buildings. Borrowing from Pagano and Bowman (2000), vacant lots are simply "unused or abandoned land" (2).

The U.S. General Accounting Office defines abandoned buildings as having been vacant for more than two years (Accordino and Johnson 2000). However, to some researchers a house is considered abandoned when it "is declared to be imminently dangerous" by local officials because it became structurally unsound due to neglect (Hillier et al. 2003, 94). The most widely cited definition of "brownfield" comes from the latest federal law governing the field: "real property, the expansion, redevelopment, or reuse of which may be complicated by the presence or potential presence of a hazardous substance, pollutant, or contaminant" (PL 107–118 [HR 2869, signed January 11, 2002]).

Comparing the Land Use Types

Table 1.1 summarizes the major distinctions among the five classifications of unoccupied urban land use types. Using three dimensions—improvements, on-site negative impacts, and off-site negative impacts—the table demonstrates how each concept is useful for different circumstance.

Table 1.1

Typology of Unoccupied Urban Land Uses

Name	Improvements	On-site negative impacts	Off-site negative impacts
HI-TOADS	Improved or unimproved	Crime, dumping, fire, disease, aesthetics, or environmental damage	Property values impacted greater than ¼ mile away
TOADS	Improved or unimproved	Crime, dumping, fire, disease, aesthetics, or environmental damage	Property values impacted up to ¼ mile away
Brownfields	Improved or unimproved	Crime, dumping, fire, disease, aesthetics, or environmental damage	Wide possible range of impacts
Vacant lots	Unimproved	Crime, dumping, disease, aesthetics, or environmental damage	Wide possible range of impacts
Abandoned buildings	Improved	Crime, fire, disease, aesthetics, or environmental damage	Wide possible range of impacts

Whether a property is improved or not affects the nature of on-site impacts. For example, unimproved properties in urban areas with limited vegetation are unlikely to generate fires, whereas properties with structures on them may. Conversely, properties without improvements are more likely to be targeted for dumping than properties with structures.

On-site negative impacts for all of the unoccupied classifications include criminal activity, dumping, fire, the spread of disease, aesthetic impacts, or environmental damage. An unoccupied property can be overrun by rodents and wild dogs, serving as a site for the spread of disease (Wallace 1989). Criminals can take over unoccupied properties as centers for their illicit commerce (Greenberg, Popper, and West 1990). Unoccupied improved properties can become sites for arsonists or contribute to the spread of accidental fires because "faulty wiring and debris create the potential for quick ignition" (Wallace 1989; Greenberg, Popper, and West 1990, 445). Unoccupied properties draw illegal dumping of household or even hazardous waste (Greenberg, Popper, and West 1990). These properties can present an aesthetic impact on their environs akin to the "broken windows" thesis advocated by Wilson and Kelling (1982), which suggests that physical signs of neglect contribute to a perception of general decay and decline in an area. Prior environmentally degrading activity on a site can worsen if a property is not maintained properly and further degradation can occur if an unoccupied property is not maintained sufficiently.

Focus on HI-TOADS

The vacant land literature is replete with questions like the one posed recently by Greenstein and Sungu-Eryilmaz (2004): "Which sites should be tackled first?" In an era of limited public resources, where should local officials target their efforts? Which are the most important sites to focus on? Powers et al. (2000) found that cities with federal brownfields grants often went after low-hanging fruit; they "looked for the projects they thought would be easiest" (32). In their study of local responses to vacant land, Bowman and Pagano (2004) developed a model whereby revenue potential, development value, and social value are aligned along three axes. Vacant land that is perceived as having low revenue, low development potential, and low social value remains out of the sights of local officials. For these properties, their "reuse will have little or no effect on the collective well-being of the community and consequently . . . interest in the vacant parcel is expected to be lukewarm at best" (165).

There remains considerable debate as to whether and how the low-hanging fruit properties, the contaminated properties, the properties attractive to development, or those with the greatest social benefits should be prioritized for development. The explosion in interest in brownfield redevelopment has meant great attention to reuse efforts, but little is known about how effective all of this interest has been in addressing the worst-of-the-worst brownfield. While prioritization may vary from place to place, HI-TOADS represent a clear and unique threat to neighborhood well-being and will be the focus of this book.

This book explores the ways that cities address HI-TOADS. Prior to examining that phenomenon, I first needed a sound method for identifying cities that were relatively likely to have neighborhoods with HI-TOADS. (1) Among large U.S. cities, which are relatively likely to have neighborhoods with HI-TOADS? After arriving at an answer to that preliminary question,

I next delved into the major questions for this book: (2) Do local planners recognize HI-TOADS as a problem in the neighborhoods of their cities? (3) In those cities that acknowledge HI-TOADS as a problem, what policies do they use to address them? (4) How successful are those policies?

In answering the preliminary question, I generated a list of cities relatively likely to have neighborhoods with HI-TOADS.[1] Are these cities largely from the traditional manufacturing belt, areas hit hardest by deindustrialization, or are they more spread out across the country? What other factors distinguish this group of likely HI-TOADS cities? Can HI-TOADS be expected only in old depressed cities or are they also present in newer booming cities?

Once the location of HI-TOADS is known, I asked: Do local planners recognize HI-TOADS as a problem in the neighborhoods of their cities? To what extent do planners and other local officials view HI-TOADS as part of the natural operations of the property market and not in need of intervention? The answers to those questions may depend, in part, upon whether planners and local officials perceive that serious neighborhood or city-wide impacts emanate from HI-TOADS. If they believe that HI-TOADS are generating severe impacts, do city officials prioritize those sites for public action?

The third question in the study is: Among those cities that acknowledge HI-TOADS as a problem, what policies do they use to address them and why? In chapter 2, I present the literature on urban policy and planning responses to HI-TOADS and organize those responses into three broad policy categories: economic development, community empowerment, and protecting public health and the environment. Do cities use these economic development, community empowerment, or public health and environment approaches? Or do they use some combination of those approaches? When it comes to comprehensive, long-term planning, do cities incorporate HI-TOADS into those efforts? What other kinds of tools or techniques do cities employ? Why do cities adopt certain approaches over others? Considering the typical age of most HI-TOADS and their inherent obsolescence, how do cities consider the challenges of historic preservation when addressing HI-TOADS?

Given the severe neighborhood impacts generated by HI-TOADS, how do the vast array of nongovernmental organizations (NGOs) at work in cities address HI-TOADS? Equally as important as what they may do, how and why are NGOs involved in HI-TOADS reuse? Likewise, what is the role of regional, state, and federal agencies in support of (or in place of) city efforts to address HI-TOADS?

The final question in the book is a question of success. How can success be measured in examining HI-TOADS reuse? When did policies lead directly to the reuse of HI-TOADS? What kinds of benefits accrue to the neighborhood or the city with a HI-TOAD site reuse? Can local government or NGO efforts be isolated for improving a neighborhood? Are larger market forces an important factor in which cities and which neighborhoods are successful? For those sites where a HI-TOAD site was successfully reused, is gentrification a problem?

Neighborhood Context

Much research in the urban studies and urban geography literature explores the complexity of neighborhood change. A few research projects have shown a link between general neighborhood well-being and the presence of HI-TOADS, demonstrating that HI-TOADS are correlated positively with indicators of distressed neighborhoods (Greenberg, Popper, and West 1990; Greenberg and Popper 1994; Bright 2003). By definition, HI-TOADS cause neighborhood-wide impacts on property values. Research in the environmental economics and appraisal literature has shown strong evidence that large, undesirable facilities such as hazardous waste sites generate externalities in the form of decreased values for proximate properties (Farber 1998; Reichert, Small, and Mohanty 1992; Page and Rabinowitz 1993; Kiel and McClain 1995). Hurd (2002) and Bible et al. (2005) found significant diminution in value 1,000 feet away from an abandoned creosote plant and a municipal landfill, respectively.

Simons and Saginor (2006a) reviewed the evidence of the impact of contamination and noxious uses on property values by closely examining 80 peer-reviewed journal articles. For operating industrial facilities, he concluded that property value losses are the greatest closest to the facility and were generally in the 15 to 30 percent range. Wider zonal impacts extended as far as several miles away and cut property values by between 4 and 8 percent. For landfills and Superfund sites, homes close to the boundary of the site experiences 10 to 25 percent losses in value, while as far as a mile away losses were still in the 4 to 8 percentage range. Simons and Saginor (2006b) also conducted a meta-analysis of 75 of these peer-reviewed studies to examine further the broader trends in property value impacts. They found that distance to a site (whether an active noxious facility or a shuttered one) was a statistically significant variable in explaining housing values across the studies. "As a property is located away from the source, the effect on price is positive and losses get smaller" (83).

Leigh and Coffin (2005) looked particularly at the property value impacts of brownfields in Atlanta, Georgia, and Cleveland, Ohio, and also found a significant spatial relationship between proximity and price. "Brownfields, both listed and potential, lower property values in Atlanta and Cleveland. The effect is strongest in the immediate vicinity (within 500 feet) of a brownfield" (276). But very large brownfields, like HI-TOADS, are known for generating some of the more severe property value impacts. Kiel and Williams's (2006) examination of the impacts of Superfund sites discovered that particularly large sites with few blue-collar workers have the highest negative price impacts on surrounding property values.

Greenberg et al. (2000) created the HI-TOADS concept based on surveys of tax assessors, where they were asked about property value impacts at four levels "(1) on the site itself, (2) within one-quarter mile of the site, (3) one-quarter to one mile, and (4) more than a mile from the site" (721). These categories were generated based on the findings described earlier, accounting for the spatial variance in expected property value impacts. That is, homes in close proximity to an undesirable facility or brownfield are expected to have some depressed property values (categories 1 and 2), and homes in a wider zone of impact should have only a minimal property value diminution (categories 3 and 4). This approach allowed for the researchers to capture a new class of brownfield, one that generated a uniquely greater degree of property value impact than the typical noxious site.

The Greenberg et al. (2000) survey results showed that one-quarter mile (1,320 feet) represented a threshold level indicating a new unique type of brownfield that had greater impact on property values. These finding are both consistent with the prior literature on property value impacts and revolutionary in how they are different. While others have studied different types of undesirable land uses, the Greenberg et al. (2000) study is the first to test the hypothesis that certain large brownfields may have even greater property value impacts than a typical brownfield site. The remainder of this book will rely on the assumption that particularly large and noxious brownfields (labeled HI-TOADS here) have a wider property value impact than the typical brownfield site. While scant evidence pinpoints the exact breadth of that impact, this study uses the quarter-mile threshold to approximate some type of sizable reach: further than an abutting property, but within the neighborhood.

HI-TOADS often are redeveloped to become something new (an office building, a park, housing). Given the significant neighborhood impacts of HI-TOADS, this book will examine how neighborhoods change when HI-TOADS are converted. However, neighborhoods change for many reasons and converted HI-TOADS are only part of any positive or negative transformation.

An important dimension of neighborhood change is the question of causality. Accordino and Johnson (2000) wrote "property abandonment has generally not been addressed as a problem; it has been viewed as a symptom, and not a cause, of urban disinvestment" (302). Burchell and Listokin (1981) argue that abandonment (and HI-TOADS by extension) are both "a symptom and a disease . . . exercising a distinct feedback mechanism which accelerates and perpetuates urban decline" (15).

HI-TOADS typically are generated due to an incompatibility of a particular land use with market demand in an area. A tannery may be abandoned by its owners as a cost-savings measure due to the exorbitant retrofitting costs to convert it for a modern in-demand land use. Why the tannery is obsolete and why the retrofitting costs are too high depend upon factors as diverse as local shopping preferences, engineering considerations, international trade patterns, national defense policy, and global labor markets. These larger social and economic forces are constantly at work not only in setting the context for the generation of HI-TOADS, but also in influencing political decision-making, and impacting neighborhood change through crime and unemployment.

In an idealized sequence of events, after identifying HI-TOADS as a problem in their community, the local government will act (but governments also direct their efforts specifically at neighborhood change). If the HI-TOADS policies are successful, the effect includes converted HI-TOADS, which themselves can then have impacts on neighborhood change. But certain kinds of neighborhood change can also affect HI-TOADS, making them more or less appealing for redevelopment.

The HI-TOADS problem is essentially a problem of neighborhood impacts. Is local government intervention in HI-TOADS the central independent variable in explaining neighborhood change? Because these relationships between government action, HI-TOADS, and neighborhood change are neither one way or static, this research does not lend itself to a simple linear thinking that can be captured in a regression analysis or even several stages of regression analysis. Modeling this phenomenon is better accomplished through a range of different research methods.

I identified U.S. cities that are relatively likely to have neighborhoods with HI-TOADS. Then I conducted interviews with local officials at these cities and I learned how they perceive HI-TOADS and about the creative and innovative strategies they implemented in addressing HI-TOADS. A series of five case studies further teased out answers to the second, third, and fourth questions asked above. This research project provides the first national study examining the effectiveness of HI-TOADS conversion efforts.

I evaluated and explored the policies cities employ when faced with HI-TOADS. This research is valuable in better understanding the ways that these policies can be used beyond addressing HI-TOADS but within the larger planning and policy environment.

This research also contributes to a larger policy debate at the local, state, and federal level about how sites are prioritized for redevelopment. This research illustrates the value in prioritizing the sites with the greatest neighborhood-wide impacts. Current federal and many state policies encourage local governments to take the easy course, rather than challenging them to deal with their HI-TOADS. The U.S. Environmental Protection Agency developed a Hazard Ranking System (HRS) to assess environmental and public health impacts of the most contaminated sites in the country and then prioritized remediation funding for those sites with high scores and in consideration of other factors. While the HRS has been criticized on a number of fronts, a comparable ranking and listing system for HI-TOADS could be useful in prioritization for public funding (National Research Council 1994).

Chapter 2 includes a discussion of what the relevant literature suggests the expected answers are to this book's questions. Chapters 3 and 4 present the results of a statistical analysis of demographic and land use data and

interviews with local officials in U.S. cities relatively likely to have neighborhoods with HI-TOADS. Chapters 5 through 9 each present a case study
of a U.S. city identified as important based upon the telephone interviews,
further answering my second, third, and fourth questions. This diverse
group of cities varies in size, region, and levels of success at addressing HI-
TOADS. They are Trenton, New Jersey, Pittsburgh, Pennsylvania, New
Bedford, Massachusetts, Youngstown, Ohio, and Richmond, Virginia.
Chapter 10 summarizes the book's findings and presents the implications of
the study for research, theory, and practice. By providing a road map to
tackling HI-TOADS, this final chapter will detail the essential lessons
learned for those engaged in the struggle to improve urban environments
and create sustainable communities.

[2]

Research, Writing, and Thinking
on HI-TOADS

Over the last few decades, much has been written about HI-TOADS and the broader questions posed in this book. An older and deeper literature on the topic of neighborhood change is also useful in introducing this study. In this chapter, I review the research, writing, and thinking that will form the basis for this study of HI-TOADS in U.S. cities.

1. Among large U.S. cities, which are relatively likely to have neighborhoods with HI-TOADS?

Bluestone and Harrison's (1982) landmark study defined deindustrialization as the "widespread, systematic disinvestment in the nation's basic productive capacity" (6). This fundamental change in economic activity "left behind . . . shuttered factories, displaced workers, and a newly emerging group of ghost towns" (6). Called "one of the major transformations of the twentieth century," deindustrialization is key to understanding where HI-TOADS appear. Cities afflicted by deindustrialization lose businesses and high-paying, low-skill factory jobs. Many of those fleeing businesses fail to make proper arrangements for the real estate they leave behind. Keenan, Lowe, and Spencer (1999) summarized the past three decade's research examining abandoned and vacant properties and concluded: "underneath most of these cases lies the problem of declining, if not outright collapse, of major industrial sectors" (706).

In his case study of Oakland, California, Self (2003) argues that deindustrialization can best be understood in terms of the "spatial dynamics of industrial restructuring" (160). The recruitment of new industry to suburban Alameda County played an important role in the decline of industry in Oakland. Cities that suffered the impacts of deindustrialization and had heavy losses in manufacturing would be expected to have a large number of HI-TOADS. However, some cities, such as Boston and Philadelphia, have emerged from deindustrialization and devised new, modern avenues for growth and would not be expected to have many HI-TOADS.

The forces of deindustrialization are connected to a larger process of spatial restructuring known as globalization. Castells (2000) has studied the concept of a networked society and discovered that it generates a new logic of industrial location. Production processes are segregated into geographically distinct areas corresponding primarily to labor markets (that is, financial accounting in center city, sales in suburbs, unskilled assembly in the Third World). Sassen (1991) also has written about the effect of globalization on major cities. According to Sassen, processes of globalization have meant that firms have spread their activities and production processes throughout the world (to take advantage of labor and regulatory cost savings). Her analysis concludes with the identification of London, New York, and Tokyo as global cities to serve as the command centers for this decentralized world. Sassen paints a more bleak picture for the thousands of smaller and medium-sized cities that play less-significant roles in the global cities hierarchy. In those cities left outside of this new global circuit of capital, much of their once-prized industrial lands today lay fallow.

U.S. cities in the traditional manufacturing belt, which stretches from Maine to Saint Louis, have been hit hardest by deindustrialization (White, Foscue, and McKnight 1964; Bluestone and Harrison 1982). The close connection between declining industry and the generation of HI-TOADS led me to expect that there would be many HI-TOADS in the traditional manufacturing-belt cities.

The next major factor in helping to explain the appearance of brownfields is what I call "metro shifts." Mass migration out of rural areas during the nineteenth century was followed in the mid-twentieth century by mass migration out of cities into suburbs (Clark 1989). When a city loses population, demand for property likewise declines and further contributes to a climate that encourages abandonment. When the economics of maintaining a property up to code and meeting environmental obligations exceeds potential revenue generation, owners often will abandon their properties and seek to avoid liability (Bowman and Pagano 2004). Therefore, HI-TOADS are expected to be found in these vacated places, which over the last several decades have been largely in areas of cities where industry concentrated.

The urban historians tie these metro shifts to three major federal policy interventions: postwar mortgage programs, urban highway building, and urban renewal. Shortly after World War II, federal support for new housing construction and mortgage subsidies led to a veritable building boom throughout the nation outside traditional urban centers (Jackson 1985). In the 1950s, the federally subsidized development of highways through cities has been widely seen as hastening the exodus of middle-class and primarily white residents from U.S. cities (Byrum 1992). Likewise, the urban renewal efforts that occurred in the 1950s and continued into the 1960s, while possibly designed with good intentions, resulted in decimated neighborhoods and are seen by some researchers as a key contributor to urban decline (Gans 1962). Together, these federal interventions, as well as local and state efforts to stem population loss, are blamed by many for creating the conditions that exist today in many poor, inner-city neighborhoods.

Recent renewed interest in downtown and inner-cities development suggests that abandoned property located in very close proximity to central business districts or popular downtown neighborhoods would be quite valuable (Schwab and Middendorff 2003). As a result, these properties generally are quick to be developed and HI-TOADS would be uncommon in such locales (Schwab and Middendorff 2003).

The final major force at work in creating HI-TOADS is a broader regulatory framework that makes property reuse expensive and difficult. The imposition on the development community over the last thirty years of an enhanced regulatory framework has made brownfields reuse challenging. Called the "ubiquitous agent in the evolution of urban form," public regulation of land use effectively limits and controls development (Platt 1996, 44). The introduction of federal and state controls on air, soil, and water pollution, wetlands and riparian protections, and hazardous materials have added to the costs of development.[1] In a market economy, these regulatory requirements can be so burdensome that they can make compliance prohibitively expensive (Colton 2003; Brachman 2004). Just as declining rents may make abandonment appealing to some, the financial burden of regulatory compliance also can have such an effect (Brachman 2004). For example, Greenberg et al. (1992) found that New Jersey's Environmental Clean-up Responsibility Act (ECRA) was perceived by some communities "as constraining the redevelopment of TOADS" (122).

Cities in states with strong environmental laws (especially limitations on property transfer) would be expected to host many HI-TOADS. States and localities that relax or expedite development conversely would be expected to host fewer HI-TOADS.

This section summarized the three major forces at work in the generation of HI-TOADS. What appears unclear is why the market does not adjust sufficiently to allow for HI-TOADS to be reused and what kinds of public intervention are most effective.

2. Do planners recognize HI-TOADS as a problem in the neighborhoods of their city?

There are two distinct bodies of thought on neighborhood change from which planners draw when confronted with HI-TOADS. Under both the neighborhood life-cycle theories, and the alternative neighborhood-change theories, planners either may see a need to intervene or expect that intervention is imprudent. When planners believe that there is a need to intervene, do they also see HI-TOADS as a priority? Or, do planners believe that their city governments should not play any role in intervention in HI-TOADS?

By viewing neighborhood change in terms of a life cycle, one theory posits that places grow and die in a way analogous to the human body: "the constant cycle of birth, life, and death is inevitable in both . . ." (U.S. Federal Home Loan Bank Board 1940, 3). Hoover and Vernon (1962) described five stages: new development, transition, downgrading, thinning-out, and renewal. The Real Estate Research Corporation (1975) outlined five similar steps along a continuum: healthy, incipient decline, clearly declining, accelerating decline, and abandoned.

Neighborhood life-cycle theory was developed in order to better understand and to rationalize the declining city. Many writing on the topic ended by identifying planning and policy interventions to either arrest or reverse this "natural" process (Bradbury, Downs, and Small 1982). Their stated goal was to help these devastated places be reborn or to prevent deterioration of existing stable neighborhoods. Neighborhood life-cycle theory has been tremendously influential in urban policy and planning of most U.S. cities, but has been subject to insightful critique (Metzger 2000).

Blakely (1994) draws on neighborhood life-cycle theory in advocating public intervention through monetary investments in vacant land in the belief that they will arrest the dying process. Described as redevelopment or revitalization, this approach often is top-down and uses forced relocation to achieve its objectives. A notorious example of this approach is the Boston Redevelopment Authority's urban renewal program in the West End of Boston (Gans 1962; Teaford 2000). More recently, the City of New London, Connecticut, won a Supreme Court decision that will allow them to move forward with the condemnation of 64 privately owned homes in order to allow a large corporation to expand (Langdon 2005a; Salzman and Mansnerus 2005). The 2005 *Kelo v. City of New London*, 125 S. Ct. 2655 case has generated a groundswell of popular sentiment against eminent domain for the purposes of economic development, provoked a rash of new state laws, and stirred up public protests against government taking of private property for economic development (Egan 2005). I expect that planners will respond to this "storm of . . . protest" by seeing a diminished role for eminent domain in the redevelopment of HI-TOADS (Egan 2005, 1).

The dominant interpretation of neighborhood life-cycle theory is that public investment is needed to stop an out-of-control process. This view of neighborhood change fails to account for a scenario where a city loses population but does so without suffering the expected accompanying blight. Neighborhood life-cycle theory conflates population loss with blight and neighborhood deterioration. Rather than look for ways to manage population loss through a HI-TOADS strategy so that blight does not occur, the theory only allows for the neighborhood to be seen as growing or declining, alive or dead.

The extent that a planner, drawing on neighborhood life-cycle theory, will perceive a HI-TOADS problem depends on how far along the site's neighborhood is in its life cycle. If the planner believes that a neighborhood is essentially healthy, then she will be less likely to intervene and the HI-TOADS will be expected to regenerate themselves with the help of the hidden hand of the marketplace. However, if the planner believes that the neighborhood is essentially dead, then she will perceive the HI-TOADS as a problem and will seek to intervene. But what of the transitional neighborhoods, neither thriving nor dying. Will planners perceive HI-TOADS to be a problem there? More importantly, how will they act upon those perceptions?

According to Metzger (2000), the future of a city depends not on its stage in a "natural" life cycle, "but on whether residents had access to financial resources within an environment of community control" (7). Metzger is drawing on a body of critical theory that rejects the modernist notions of advance and retreat, of growth and decline. Beauregard (2003) also explores this dialectic in examining the discourse of "urban decline." He finds that urban decline was incorporated into a socially constructed story of the rise of suburbia and fall of the city—a fictional account that has been reified into the public consciousness through oral and written communication.

A postmodern notion of neighborhood change, like that presented by Dear and Flusty (1998), escapes this grand narrative and allows the details of each city, each neighborhood, and each block to speak for themselves. Mitchell (2002) also contributes to this alternative theory in his account of planning in Egypt. He shows how it was the "informal, clandestine, and unreported" activities of society that determined outcomes, not the "fabricated" script developed by Western colonizers. An understanding of urban decline as a disaggregated, finely complex phenomenon is possible under this alternative theoretical framework. This alternative theory of neighborhood change allows planners to be cognizant of urban problems, but to avoid the all-encompassing "urban decline" discourse. Such an unshackling from the structures of urban decline opens up the possibility for the planner to work toward HI-TOADS reuse.

A planner drawing on this alternative theoretical framework may attempt creative intervention as described above or may altogether avoid action. Hoch (1996) suggests that a consequence of postmodern planning practice is that a sense of hopelessness may infect the planner because all interventions are somehow intertwined with the forces of power. The planner who embraces alternative neighborhood-change theories may be reluctant to force a label on a property (such as a HI-TOAD site) or timid about his own agency in manipulating power relations in an affected neighborhood.

Theory aside, a significant body of research has examined the perceptions of planners concerning brownfields and HI-TOADS. Greenberg, Popper, and West (1990) conducted interviews with local planning and health officials in 14 of the 15 largest U.S. cities, and then Greenberg and Popper (1994) administered a mail survey to health officials in "distressed northeastern and midwestern cities with between 25,000 and 500,000 residents" (25). A third study looked at medium-sized cities in New Jersey and the latest examined all municipalities in New Jersey through a mail survey (Greenberg et al. 1992; Greenberg et al. 2000). Using a more general definition of TOADS based closely upon the definition provided above for vacant lots and abandoned buildings, the 1990 study found that more than half of big U.S. cities saw TOADS as a problem in their city and specifically identified fire hazards, toxic waste, drugs, and homelessness as directly exacerbated by TOADS.

The 1992 study of New Jersey cities with populations between 20,000 and 315,000 found that of 21 communities in the sample, 10 "reported that TOADS were a serious land use problem" (Greenberg et al. 1992, 119). One city official reported, "These areas disrupt the fabric of the cities" (119). The 1993 Northeast and Midwest mail survey further confirmed that local officials are largely aware of TOADS and their impacts. The most recent, a statewide New Jersey study, introduced the more refined TOADS definition that considered impacts on property values and transactions beyond the property's borders (Greenberg et al. 2000). Under this definition, 46 of 450 New Jersey municipalities responded that they had TOADS in their community, while 15 responded that they had HI-TOADS. Those HI-TOADS communities were mostly large (mean population of 46,007 compared with a mean population of 13,658 for municipalities statewide in 1990), poor ($15,348 per capita income compared with a per capita income statewide of $18,714 in 1989), and with a relatively high non-white population (28 percent non-white compared with a statewide non-white population of 21 percent in 1990). My expectation is that planners in large, poor cities with high non-white populations will acknowledge that there are TOADS in their communities and that those TOADS represent a problem.

In their survey of the 200 most populous cities in the United States, Accordino and Johnson (2000) found that local officials in western cities did not perceive vacant and abandoned property to be "particularly serious, overall" (305).[2] Only 21 percent of western officials agreed that vacant and abandoned property represented a "big problem" or the "biggest problem," while 67 percent of officials in northeastern cities felt that it did (305). A plurality of officials in the South and Midwest agreed that vacant and abandoned property was a problem. This distribution is likely caused by a variation in the scope of the actual abandonment problem among the regions. However, officials in different regions also may perceive the same phenomenon differently and assess its severity differently. Because there is no literature on differing perceptions on abandonment among regions where actual abandonment levels were controlled for, there is a need to assess this variation in the study. Notwithstanding differences in perceptions, I expect in this study that the most severe HI-TOADS problems will be reported in the Northeast, then the Midwest and South, and lastly the West. These four U.S. Census regions organize much of the inquiry presented in this book.

The closure of major plants in the second half of the twentieth century was examined closely by Mayer and Greenberg (2001) in a study of small and medium-sized U.S. communities. They concluded that strong leadership was the single most important variable in explaining the ability of the community to rebound from the devastation of a major plant closure. In this book, I likewise will expect that planners in cities with strong leadership will be most likely to recognize that there are HI-TOADS in their community and be prepared to act.

3. What do cities do to address HI-TOADS?

Cities utilize three main classes of policies when faced with HI-TOADS: economic development, community empowerment, and protection of public health and the environment.[3] To be clear, there is not necessarily any spatial-temporal consistency to the application of these policies: Some cities use some policies sometimes in some places, while using other policies other times in other places. All three policies could be utilized by a single city in a single neighborhood all at the same time.

Something as simple as a definition of "planning" is in dispute in the literature. Even with clarity on what constitutes planning, few scholars can agree on who is a planner. This book is about both the act of planning and the role of professional planners.

According to the dictionary, planning means " . . . to devise a scheme for doing, making, or arranging" (Guralnik 1986, 1088). Early professional planning took that straightforward verb and applied it to public sector policy-making. In 1935, Charles Merriam said that planning is an "organized effort to utilize social intelligence in the determination of national policies . . . based upon fundamental facts . . . thoroughly analyzed . . . [to] look forward to the determination of longtime policies" (as quoted in Boyer 1983, 204). Planning became more locally and regionally oriented during the twentieth century, and Campbell and Fainstein (2001) articulated that new definition well: "the central task of planners is serving the public interest in cities, suburbs, and the countryside" (13).

Lindblom (1959) defined the practice of planning as "the science of muddling through," where professionals seek to organize their work around pursuing key goals, but spend most of their time facing smaller challenges and work to address each of these. This incremental view of planning puts the daily struggle of policy and politics to the fore of the planner's agenda, with less of a commitment to long-term plans.

Friedmann's (1987) contribution to defining planning had three dimensions: "Planning attempts to link scientific and technical knowledge to actions in the public domain; Planning attempts to link scientific and technical knowledge to processes of societal guidance; Planning attempts to link scientific and technical knowledge to processes of social transformation" (38). Here, planning is work that links various types of knowledge to actions and processes. Under Friedmann's definition, there is much room for planning outside the types of activities delineated by the American Planning Association, American Institute of Certified Planners, or the Association of Collegiate Schools of Planning (that is, land use, environment, transportation, economic development, and the like). Therefore, in the interest of keeping this research focused on urban redevelopment practice, I proffer the following definition based upon my review of the literature and my own experience as a professional planner: Planning is a space-based, future-oriented activity involving the setting of goals and means to achieve those goals, and involving some level of public participation. Included in this definition is all activity focused on a site, city, town, watershed, region, or activities in the substantive areas of transportation, housing, or economic development, or other space-based areas.

In order for a government body to be engaged in planning, I add a corollary to my definition: that such goals and means must be codified formally. Without written adoption of goals, means, and forward thinking, such activities are not government planning. Activities that do not qualify under my definition of government planning are worthy of study, but will be considered as quasi-planning activities.

In addition, for the purposes of this book, any professional engaged in activities under my definition of planning is a planner. This title can be applied to professionals working in any agency or office. For example, an economic development specialist working within a Brownfields Office but engaged in the activities under my definition of planning is a planner.

As was demonstrated in Rodwin and Sanyal's 2000 book, *The Profession of City Planning: Changes, Images, and Challenges 1950–2000*, planners come in many stripes and are engaged in many different kinds of substantive areas (education, urban design, community organizing, economic development, housing, and more). I expect that, due to the nature of HI-TOADS, planners will be involved heavily, that the practice of planning will be employed, and that local governments will do planning (that is, they will codify their goals and objectives stemming from planning activities).

While variation does exist among locales, urban regime theory provides for a useful umbrella of ideas to explain how cities plan. Stone (1989) wrote that a regime is "the informal arrangements by which public and private interests function together in order to be able to make and carry out governing decisions" (6). This collaboration for common interests is hardly the ideal of a totally participatory democracy. Yet it is not quite so distasteful as the autocratic planning exemplified by Robert Moses (Caro 1974). Under urban regime theory, public sector planners work closely with private developers, civic associations, state and federal regulators, and politicians to carry out projects.

Critics have pointed out the "localist" flaws of urban regime theory. Ward (1996) suggests that external political, social, and economic conditions are not accounted for sufficiently when the scale of analysis is the local regime. Sites (1997) argues that by downplaying those larger forces, urban regime theory masks the important role that "social movements or community mobilization [play] in urban politics" (551). Consequently, the urban regime theorist explains the way a city operates without fully accounting for the work of grassroots organizations and "underestimates the obstacles to genuinely progressive, community-oriented urban development" (537).

For this study, I expect that regimes will play an important role in responding to or failing to respond to HI-TOADS. However, I also expect that the work of civic associations (including community groups) functioning outside powerful regimes will be important to understanding the holistic local response to HI-TOADS.

The first important tool that cities have to address HI-TOADS is economic development. Predicated on the belief that "the most depressed and desperate communities can be revived through careful planning and judicious leadership," economic development policies offer a range of incentives, grants, loans, and technical assistance packages to firms to support urban revival (Blakely 2000, 285). Recognizing the ill effects that unregulated markets can have on cities, economic development is the local version of Adam Smith's hidden hand of government. By intervening in basic supply-and-demand equations, economic development policies make urban investment more palatable to firms. Economic development efforts also seek to generate a more business investment–friendly environment in a given locale by enhancing its workforce and improving its unique locational assets (Blakely 2000).

Economic development is grounded largely in neighborhood life-cycle theory and responds to decline by attempting to arrest the decline and reverse the process. Economic development uses redevelopment and revitalization tools to "stop" a whole urban decline process and instead make it grow. Growth and decline are seen as large processes connected to root causes such as deindustrialization, decentralization, and disinvestment from inner cities. Economic development policies attack the root causes of population loss, make efforts to actively reverse the loss, and address the impacts.

Key to economic development are tools such as development incentives, the use of eminent domain to seize blighted properties, and aggressive modifications to land use regulations (Bailey 2004). These are the most common tools available to communities seeking to address HI-TOADS.

After their abandonment, many brownfields or other abandoned properties go into public receivership due to tax delinquency (Davis 2002). Eminent domain is not necessarily useful in converting a HI-TOAD site itself into a new use, but rather serves as part of an effort to assemble larger viable parcels of land for redevelopment (including the HI-TOAD site). For example, the redevelopment of an abandoned manufacturing facility may be financially viable only if done in conjunction with the reuse of an abutting parking lot. In such a case, a city may seek to use powers of condemnation to acquire the parking lot in order to facilitate redevelopment of the manufacturing facility.

Community empowerment is a second tool available to address HI-TOADS reuse and includes a set of policies that cities may implement with

the express purpose of putting local residents and community-based organizations (CBOs) in a position to plan for and effect change in their
neighborhood. Grounded in the alternative neighborhood-change theories, community empowerment policies do not label neighborhoods as
growing or dying, but rather position local organizations and residents at
the center of managing and at least partially directing whatever change is
occurring within a given neighborhood.

Examples of city policies that fall within the community empowerment
policy set include financial support in the form of grants and low-interest
loans to CBOs, technical support for plan-making to CBOs, open public-
planning processes aimed to elicit residents' opinions about their neighborhoods, and development of standing task forces or committees with
significant representation of community members. Successful community
empowerment policies generate neighborhood-based, citizen-driven processes whereby the people who live and work in a place are chiefly respon-
sibly for identifying problems, devising a plan to address those problems,
and implementing the plan (Thomas and Grigsby 2000). Not only is the
process driven by residents, but the aim of any improvements is "for the
benefit of the residents [of that neighborhood]" (Dewar and Deitrick 2004).

The geographically specific location of community-based organizations
limits the extent to which they are able to address larger forces such as
deindustrialization, decentralization, and inner-city disinvestment. (Although, with the help of federal laws like the Community Reinvestment
Act, community-based organizations have grown in their effectiveness to
address inner-city disinvestment.) For a city looking to address its HI-
TOADS problems, community empowerment is only one component of a
multifaceted strategy that also addresses wider challenges like regional fiscal policy.

Each of the three sets of policies described in this chapter represent a dimension of how cities address urban redevelopment. If the city's goals are
to enhance employment opportunities and tax-revenue generation, then it
will adopt economic development policies. If the city's goal is to empower
its citizens to look closely at their neighborhoods and chart a course for addressing change, then it will adopt community empowerment policies. If
the city's goals are to shield residents from toxic conditions or improve
wetlands performance, then it will adopt policies that protect public health
and the environment (henceforth, healthy environments). However, it is

entirely possible that a city would choose to build its economy, involve communities, and protect public health and the environment. This book explores whether cities do just that.

Healthy environment policies are not grounded in alternative neighborhood-change theories. Healthy environment policies focus on risks and hazards to individuals and seek to address those directly, without depending on a grand narrative of urban decline for justification. No matter if a neighborhood is comprised of wealthy people or poor people, powerful people or disenfranchised people, cities implement healthy environment policies to address the harms that afflict people. Alternative neighborhood-change theories posit that public-sector intervention in neighborhoods is justifiable only through a community process that focuses on improvements in local people's lives. Cities implement healthy environment policies to improve the environments where people live, a hallmark of alternative neighborhood-change theories. These policies also can stimulate economic development by making the neighborhood more attractive to investors.

Healthy environment policies focus on mitigating or preventing risks to people and environmental resources. Healthy environment policies are used in urban redevelopment where hazardous materials releases threaten water quality, when a derelict building threatens to collapse and cause harm to nearby pedestrians, or when airborne contaminates pose a risk to workers at an industrial plant.

For each set of policies, a certain level of political support is generated for a city to act in addressing HI-TOADS. New employment and increased tax revenues that accompany economic development are clearly valuable political points for a city administration to score. Community empowerment policies are popular with voters if they feel that their city leaders are listening to them and willing to support their agendas. Healthy environment policies can be attractive to city policy makers, especially if there is evidence that their neighborhoods have been polluted.

Under the rubric of sustainable development, healthy environment policies have become increasingly popular. Portney's (2002) study of 24 U.S. cities showed broad engagement with policies to promote sustainable development, in particular those that protected the environment. City policies to protect public health are as old as city governance itself (Hall 2000). From the first sewer systems to the imposition of constraints on industrial pollution, protecting public health has always been an important core goal of city policy-making (Hall 2000). Today, when cities face the challenges of distressed neighborhoods, they address a wide range of public health challenges (Greenberg 1996). HI-TOADS generate their own set of risks to public health and in this book, I describe examples of cities directly addressing those risks.

According to Bowman and Pagano (2004), cities have health and safety policies that they employ to address abandoned properties that pose a risk. Whether burdened by dumping on the property that causes public health threats to nearby residents or unstable structures that threaten to collapse on passing pedestrians, cities can choose to act to contain or mitigate against these dangers under the authority of their health and safety regulations.

Bowman and Pagano (2004) also found that some cities try to prevent a property from becoming derelict using a registration fee for abandoned buildings or an abandonment tax. These policy tools impose additional taxes and fees on property owners if they abandon their property. The policies encourage owners to keep their properties in a secured, partially occupied status until a reuse option is possible. This transitional state is called "mothballing" and sometimes can be a problem in itself. In Greenberg, Downton, and Mayer's (2003) survey of local officials in 68 distressed New Jersey communities, a third reported that mothballed properties did represent a problem in their community. Despite the externalities generated by mothballing, Cohen (2001) recommended such an approach to the City of Baltimore so that properties "could be stabilized at relatively low cost without being an eyesore" (434).

The literature is inconclusive with respect to what policies cities use when addressing HI-TOADS. The economic development approach would favor aggressive policies utilizing development incentives and the use of eminent domain to bring jobs back to an abandoned site through a new development project. The community empowerment approach would favor community-benefiting activities at a HI-TOAD site, including community facilities. Under the healthy environment approach, the goal would be amelioration of public health and environmental impacts of the HI-TOAD site, and an attempt to make the neighborhood more attractive. The literature offers little in the way of answers about which cities would employ which approaches or even if they develop policy responses beyond these three classes of options. In this book, I will look to classify cities' actions into each of the three categories of economic development, community empowerment, or healthy environments.

4. How successful are cities at addressing HI-TOADS?

The final question of this book addresses a large gap in the literature: how successful cities are in their efforts to address HI-TOADS. In Bright's (2000) national investigation into inner-city revitalization efforts, she defined a successful intervention as one that leads to a measurable improvement in the quality of life of existing residents. She focused on safety, services, shelter, and social capital as indicators of quality of life, explicitly rejecting wealth indicators like income or property values. This approach allowed her to isolate those revitalization efforts that improved residents' lives. This book's focus on public policy and planning responses to the worst-of-the-worst abandoned sites within a neighborhood will require a different definition of success.[4]

A recent study looking at redevelopment initiatives in Texas found that relatively small public investments in redevelopment generated substantial positive economic impacts for the neighborhood and beyond (Bright 2003). However, the study also found that, in some cases, the direct fiscal impact on a single funding agency was negative in the short term, where "public sector costs exceeded revenues" (Bright 2003, 28). Another key finding of the Texas study was that for abandoned properties located in high-demand markets, local governments had the biggest impacts on successful reuse when they worked toward eliminating barriers to reuse such as title problems, site clearance costs, or infrastructure needs.

Wernstedt (2004) reviewed fourteen major studies of community impacts of brownfields reuse and found substantial benefits in terms of the number of new jobs, property value enhancement, and less tangible improvements. One such study found that the EPA's Brownfields Pilot program resulted in 457 brownfields redevelopment projects yielding 17,000 jobs (U.S. Environmental Protection Agency 2000). Another showed a median of 250 jobs and $1 million in tax revenues for each redeveloped brownfield site (U.S. Environmental Protection Agency 2003). A New Jersey study extrapolated survey results collected on 83 sites to estimate that between 17,000 and 60,000 new jobs and between $55 and $287 million in new tax revenues could be expected from brownfields redevelopment in the state (Miller et al. 2000).

In the literature, a number of studies examined successful revitalization efforts. In her 2000 study of thirteen U.S. cities, Bright investigated examples of successful revitalization efforts by culling the literature for examples of quality of life improvements to residents. She distilled nine procedural strategies employed in each city that led to quality of life improvements. These strategies included greater empowerment of residents to lead revitalization, a regional approach, involvement of the private sector, and effective tracking and reuse of TOADS.

A special city program in Pawtucket, Rhode Island, reduced the number of parcels of vacant and abandoned property by 90 percent and "increased self-esteem of home owners and . . . improved quality of life in these residential neighborhoods" (Doyle 2001, 16). The blight removal "created a healthier living environment for residents in the neighborhoods" (Doyle 2001, 16). Brownfields redevelopment programs in Portland, Oregon, and Trenton, New Jersey, are two examples of where the reuse of derelict property resulted in the creation of much-needed affordable housing, thus improving residents' quality of life (Bartsch 2002).

In environmental terms, any reuse of a contaminated HI-TOAD site has the potential to have some positive environmental impact. The brownfields literature is replete with examples of positive environmental impacts occurring with the reuse or redevelopment of abandoned sites with real or perceived contamination (Bartsch and Collaton 1997; Powers et al. 2000). Harrison and Davies (2002) interviewed urban ecologists in London and found that brownfields were often also sites of high ecological quality and reuse further enhanced that value. Included in Wernstedt's (2004) review is reference to a study by Deason, Sherk, and Carroll (2001) that estimated that for every acre of brownfields redeveloped, development was steered away from a median of 1.7 acres of greenfields. De Sousa (2003) demonstrated the potential accrual of positive benefits to ecological habitat due to brownfields reuse.

Several studies, some using hedonic price modeling, have explored the way that abandoned housing has a contagion effect, spreading further abandonment (Wilson and Margulis 1994; Kiefer 1980; Odland and Balzer 1978; Wallace 1989). These studies have shown how abandonment can result in greater abandonment. Through that logic, reusing and restoring abandoned buildings (or HI-TOADS) would arrest the spread of blight.

There is some anecdotal evidence that this happens, but no systematic studies have tested the theory.

Market strength is seen by many as a major driver in the success of brownfields redevelopments, and I likewise expect it to determine the success of HI-TOADS conversion (Leigh and Coffin 2005; Leigh and Coffin 2002; Bartsch and Collaton 1997). Some are pessimistic, like Brachman (2004), who believes that there may be a "brownfields 'glass-ceiling' . . . whereby some sites may never be redeveloped" (87). I stake out a more optimistic position and make the heart of this book an investigation into how local action can successfully overcome what the market says cannot be done. Other variables such as level of contamination, size, location, and type of HI-TOADS likewise are expected to determine outcomes. A 100-acre, moderately contaminated HI-TOAD site in an excellent location likely is easier to address than a heavily contaminated, 1,000-acre, remote one. I would expect that institutional factors like a city administration's ties to key regional and national power brokers, its track record for getting grant monies, and whether its state or county has an aggressive economic development or brownfields program would partially dictate the success of HI-TOADS conversion.

[3]

Local Officials and Their Attitudes
toward HI-TOADS

While a truly comprehensive analysis of tens of thousands of cities and towns across the country would be interesting, given limited resources, this project sought to find only those cities that are big enough to have several distinct neighborhoods and that are likely to have at least a few HI-TOADS in different neighborhoods.

Methods

In order to answer the first question posed in chapter 1—among large U.S. cities, which are relatively likely to have neighborhoods with HI-TOADS—I conducted a multivariate statistical analysis of two data sets in order to arrive at a list of cities for further study. In conducting interviews with these HI-TOADS–rich cities, I asked:

2. Do planners recognize HI-TOADS as a problem in the neighborhoods of their cities?
3. Of those cities that acknowledge HI-TOADS as a problem, what policies do they use to address them?
4. How successful are those policies?

The first step in this study was to find cities relatively likely to have neighborhoods with HI-TOADS. To begin with, I constructed a sampling frame that I drew on for the subsequent telephone interviews and case studies. Because HI-TOADS are not present in every city, I needed to create a narrow enough sampling frame of cities so that my interviews (chapters 3 and 4) and case studies (chapters 5 to 9) include 30 cities that have a high probability of having multiple neighborhoods with HI-TOADS.

The sampling frame included all U.S. cities with populations greater than 100,000 in 1970 (a total of 153). From that sampling frame, I conducted statistical analysis to select a smaller sample of 30 cities relatively likely to have neighborhoods with HI-TOADS. Through this method, those cities that for historical or structural reasons may never have had HI-TOADS (for example, newer Sunbelt cities) are not included in the sample used to answer the second, third, and fourth questions of this book.

From this larger sampling frame, I then selected the cities that ranked highest on the list. This sample of cities offered two possibilities for studying the impact of planners' awareness of HI-TOADS and their policy intervention. First, among the neighborhoods in each city, I was able to compare the efficacy of local intervention (or lack thereof), contrasting HI-TOADS neighborhoods where a range of interventions (from nothing to active) occurred. Secondly, among the 30 cities, I compared those that were relatively successful overall and those that were not.

In order to produce the list of cities likely to have neighborhoods with HI-TOADS, I examined two datasets: (1) numbers of abandoned building in U.S. cities with populations greater than 100,000 in 1995 drawn from a survey conducted by Bowman and Pagano (2004) and (2) socio-economic data from the U.S. Census. Prior research has shown that the presence of TOADS and HI-TOADS is associated strongly with the existence of vacant and abandoned property (Greenberg et al. 2000). Therefore, I first compiled the results of the Bowman and Pagano (2004) survey and ranked the cities in terms of per capita number of abandoned buildings.

Then, I examined U.S. Census socio-economic data. Previous research has established links between a variety of socio-economic characteristics and the appearance of TOADS and HI-TOADS: housing value, housing vacancy, housing tenure, economic status, educational attainment, and population loss (Hillier et al. 2003; Wilson and Margulis 1994; Greenberg et al. 2000). In addition, I examined the percentage change in manufacturing employment over the last 30 years. I conducted cluster and factor analysis on these variables. Cluster and factor analysis are data-reduction techniques that allowed me to extract the essential information offered by the variables to aid me in understanding which cities are relatively likely to have neighborhoods with HI-TOADS. Based upon an earlier unpublished pilot study, a strong factor emerged indicating a close correlation among the housing, manufacturing, and socio-economic variables identified above. From this single strong factor, I examined each city's factor scores. This analysis produced a ranking of the cities, generated through factor scores, that are relatively likely to have neighborhoods with HI-TOADS. I combined the results of the analyses, as illustrated in figure 3.1, to rank all of the cities in the datasets.

See Appendix A for a detailed discussion of the implementation of this approach.

The question of which large U.S. cities are likely to have neighborhoods with HI-TOADS is answered in Appendix B. The list of cities easily could be mistaken for a lot of "worst of" lists: worst schools, worst crime, or worst tourist destinations. The cities at the top of my list are largely from the manufacturing belt (the Northeast, New England, Midwest, and the Middle Atlantic) and portions of the old South, cities well known for their decline and a legacy of abandoned heavy industry.

My methodology did more than compare cities in subjective terms, it translated five strong indicators of HI-TOADS into a single factor for which the cities were objectively ranked.[1] The remainder of the study will rely heavily on data collected from cities identified in this chapter. The benefit of this approach is that I will not spend time collecting data in Stamford, Connecticut, Raleigh, North Carolina, or Austin, Texas, cities ranked 138, 143, and 149, respectively. With such low rankings, my analysis has found that they are cities with statistically low probability of having neighborhoods with HI-TOADS.

Next, I conducted telephone interviews with planners and other local officials at the cities that ranked the highest using the Stage One analysis. These interviews were conducted during the fall of 2005. Officials at some of the cities were unwilling to cooperate. As required by my protocol, I dropped down further on the list of cities to number 31 in order to complete the interviews.[2]

The planner is typically the official agent within a city who can speak on issues of land use and development. Brownfields coordinators, economic development officials, mayors, and city managers are also often quite knowledgeable about land use and development activities in their city (in some cases even more so than the planner). By interviewing these city employees, I will hear representatives of the official voice of the local government. Of course, such a voice does not fully represent the opinions, values,

FIGURE 3.1. Identification of cities relatively likely to have neighborhoods with HI-TOADS.

Abandoned building data set

		Low relative likelihood to host HI-TOADS	High relative likelihood to host HI-TOADS
Socioeconomic data set	High relative likelihood to host HI-TOADS	Cell #2	Cell #1
	Low relative likelihood to host HI-TOADS	Cell #4	Cell #3

Note that analysis of each data set will break down cities into either "high relative likelihood to host HI-TOADS" or "low relative likelihood to host HI-TOADS." Interviews were conducted with cities that fell in cell #1.

and perceptions of all of the city's population, it is merely one version. However, the third stage of this research sought to corroborate that official voice through in-depth interviews with others in the community.

To ensure that at least twenty cities were interviewed, I contacted officials in the top thirty cities identified in Stage 1.[3] I reviewed each city's website and spoke to city clerks and planning department staff to find a primary point of contact. Then, using a snowball-sampling technique, I asked the primary point of contact for a referral to a second local official who was knowledgeable about HI-TOADS.

I employed a modified version of Dillman's (1978) total design method to ensure a high response rate for the interviews. Prior to each interview, I sent a letter of introduction describing the research project and inviting the addressee to participate. In the letter, I provided a written definition of HI-TOADS, as well, as the interview instrument and informed consent agreement forms (see Appendix F).[4] My protocol to secure interviews is outlined in Appendix A.

In the interviews with the local officials, I asked if there were any or had been any HI-TOADS in their community in the last ten years. If they had examples of HI-TOADS, I asked which policies, if any, were implemented to address them. I asked how successful the community's policies were in addressing its HI-TOADS, and why they were successful or why they were not. I measured success by the extent to which a known HI-TOAD site no longer meets the definitional criteria introduced earlier: real property, abandoned or largely abandoned, which is estimated to impact negatively on property values more than one-quarter mile away. The one-quarter mile designation is used as an approximation for passing a threshold of important neighborhood-wide impacts. That is, the property value diminution is not limited to a few abutting properties, but rather extends deep into the surrounding area. In my interviews, I asked about any estimated property value impacts in any direction from the site roughly more than one-quarter mile away.

Building on the Bowman and Pagano (2004) findings, I parsed out cities' policy tools to address HI-TOADS, as distinct from overall vacant-land policies. For example, I inquired if they had earmarked HI-TOADS for special development incentives, or if they had adjusted zoning and subdivision rules in HI-TOADS zones to accommodate reuse. Additionally, officials commented on the impact HI-TOADS had on their neighborhoods over the last ten years.

In examining 1970's biggest cities, this study concentrated on a sample of cities large enough that they are comprised of a number of distinct neighborhoods. Such a large sample of sites allowed me to control for market strength, size of the HI-TOADS, type of HI-TOADS (waste disposal facility, former industrial site, etc.), and institutional arrangements. I expected that, for each city interviewed, I would hear about both successes and failures, both in the past and present. Were all of the successes in certain neighborhoods and all the failures and challenges in others?[5]

Next, I conducted a statistical analysis of the data using cross-tabs to explore the relationships between independent variables such as a city's region or its form of government and dependent variables such as whether officials viewed the city as successful in addressing HI-TOADS. Because some variables are measured nominally, some ordinally, and some at the interval-ratio level of measurement, chi-square tests were used for most relationships, and when appropriate, Spearman's rank correlation was used. Given the small number of observations, I did not necessarily expect statistically significant results. However, the effort to quantify the interviews helped organize the data and guide the presentation of the results.

Participants

Of the 60 local officials solicited to participate in this study, I conducted telephone interviews with 38.[6] The other 22 officials either refused to participate or simply never returned my calls. The average interview lasted approximately one hour. Most often, I conducted the interviews in person, but some were conducted over the telephone. Most interviews were one-on-one, but on occasion, I spoke to more than one research participant simultaneously. This was only done at the request of the participants. The 38 officials are from 21 cities. With the exception of city population, those cities that did not participate in the interviews did not differ in a statistically significant way from the nine cities that were solicited for interviews yet did not participate (see table 3.1). These ten variables encompass seven broad socio-economic categories commonly used in urban planning research: age, income, education, race, poverty, employment, and housing vacancy. Those officials who

Table 3.1

Comparison between Cities that Participated in the Study and Those that Did Not

Variable	Non-participating cities (n=9)	Participating cities (n=21)	Z-score (two-tail)
Average population (2000)†	445,518	192,304	2.24**
Percent manufacturing change (1970–2000)	−53.1	−57.8	−0.24
Median household income (1999)†	$28,838	$27,606	−1.39
Percent over 65 years old (2000)	6.4	6.4	0.00
Percent African American (2000)	47.1	41.4	0.29
Percent white (2000)	45.3	48.1	−0.14
Percent high school graduates (2000)	74.0	71.0	0.17
Percent receiving public assistance (2000)	7.9	8.3	−0.04
Percent unemployment (2000)	6.6	6.1	0.05
Percent other vacant housing units (2000)	3.4	3.7	−0.03

SOURCE: All variables are from the 2000 U.S. Census unless otherwise noted.

† difference of means test; all other variables were calculated using difference of proportions test.

* $p < 0.10$, ** $p < 0.05$, *** $p < 0.01$

declined to participate were from a statistically significantly larger group of cities—twice as large, on average, as the cities where officials did participate.

This book is focused on the ways that planners participate in HI-TOADS projects. Therefore, I identified planners, in most cases, as the primary point-of-contact at each city. Surprisingly, many planners declined to participate in the interviews and instead referred me to other officials in the city. The officials to whom I was referred were always either in economic development or brownfields offices. Many of those I spoke with explained that their office is not involved in HI-TOADS. In those cases, I took extra time to try to explain the HI-TOADS concept, how it is a particularly severe type of brownfield.

When I asked the research participants what position they held in their city government, their answers fell into four categories: planners (17), economic development professionals (10), brownfields coordinators (6), and other miscellaneous city departments (3). Likewise, a plurality of the interviewees had professional planning training, with 15 holding either a master's or bachelor's degree in planning. Fourteen had graduate degrees in other disciplines, including public administration (6), business (2), landscape architecture (2), historic preservation (2), and one each in architecture, art history, English, forestry, geology, and sociology. The group was also fairly experienced, with a mean of nearly six years in their current position and over ten years with their city's government. As a whole, the group of interviewees had solid education and experience in urban planning and development, which contributes to my confidence in these results.

Are HI-TOADS a Problem to These Officials?

In each interview, I asked "Are there any HI-TOADS in your city?" This initial question subsequently led to an in-depth discussion about those HI-TOADS, the problems they cause, and the things that the city was doing about them. Appendix G includes a list of all of the HI-TOADS reported by the research participants.

Both local officials in Macon, Georgia, reported that there were no HI-TOADS in their city. One of the two local officials I spoke to in Savannah, Georgia, reported no HI-TOADS and the other reported five in the city. Otherwise, all other research participants reported at least one HI-TOAD site. While distribution of former HI-TOADS was somewhat normal (skewness of 1.164), there was a heavy right skew to the distribution of current HI-TOADS (skewness of 3.816) (see table 3.1). Pete Cammarata, Executive Director of the Erie County Industrial Development Agency, said that the City of Buffalo has 100 current HI-TOADS, the most HI-TOADS reported by any research participant (see tables 3.2 and 3.3 for descriptive statistics of the results of the interviews). The median city reported three *current* HI-TOADS and only one *former* HI-TOAD site.

   HI-TOADS described by interviewees included former landfills, manufacturing facilities, power plants, railyards, an office building, and a department store. Each site has its own complex and, in most cases, ugly history. Appendix G includes a brief summary of each HI-TOAD site, describing each site's location, acreage, prior use, and reuse efforts. Each interviewee relied heavily on the details of each HI-TOAD site in answering the questions posed. While some HI-TOADS are large and particularly noxious and others are small and seemingly benign, officials reported that they all probably bring down property values more than one-quarter mile away from their site boundaries. This commonality allows me to study closely the unique problems and challenges of the HI-TOAD site in the U.S. cities selected in the sample. The remainder of this chapter features the responses that local officials gave to questions about the causes of HI-TOADS and what neighborhood impacts HI-TOADS cause. Scattered throughout this and the following chapters are brief profiles of HI-TOADS to help illustrate key points that surfaced during the interviews.

HI-TOAD Site Profile: Drayton Street Gas Station, Savannah, Georgia

Situated "smack dab in the middle of the historic district," this classic 1920s-era gas station has been a drag on Savannah's otherwise thriving city center, according to Lise Sundrla, Executive Director of the Savannah Development and Renewal Authority. Abandoned by its owner/operator over fifty years ago, the 2,000-square-foot structure has suffered a slow death, with the roof caving in and the port cochere practically falling down. While not as large as other HI-TOADS included in this study, such as the 1,200-acre South Buffalo Industrial Area in Buffalo, New York, the Drayton Street Gas Station is adjacent to residential structures and compromises the integrity of Savannah's

Table 3.2

Descriptive Statistics of Key Independent and Dependent Variables from Interviews

Variable	Officials			Cities		
	#	%	n^*	#	%	n^*
City located in the Northeast	20	53	38	11	52	21
City located in the Midwest	6	16	38	4	19	21
City located in the South	12	32	38	6	29	21
Mean number of former HI-TOADS	—	—	—	1.62	—	21
Mean number of current HI-TOADS	—	—	—	9.14	—	21
HI-TOADS in low property value neighborhoods	24	73	33	13	65	20
HI-TOADS threaten stability	28	93	30	17	89	19
HI-TOADS contribute to crime	16	73	22	10	50	20
HI-TOADS contribute to arson	6	27	22	6	30	20
HI-TOADS contribute to dumping	10	45	22	8	40	20
HI-TOADS contribute to other impacts	7	32	22	6	30	20
City successful in addressing HI-TOADS	24	69	35	14	70	20
Successes resulting in gentrification	12	34	35	8	40	20
City embraces green policies	11	35	31	8	44	18
City received external grants	28	88	32	17	94	18
Value of grants in last year	$6,933,333	—	27	$7,226,470	—	17
Cities with strong-mayor form of government	—	—	—	15	71	21
Cities with council form of government	—	—	—	2	10	21
Cities with control board form of government	—	—	—	2	10	21

world-class historic center. While Savannah officials did not report crime associated with the gas station, one official reported that the site was a key factor depressing area property values and a visual blight on the neighborhood. Only recently has this site received the city's attention; for decades it was not seen as a city priority.

Questions about whether there are any HI-TOADS in their city provoked many local officials to talk about their city's history and how that affected whether any HI-TOADS exist today. This section provides anecdotes that may explain what the respondents believed may have caused or contributed to HI-TOADS in their communities. Mr. Cammarata, of Buffalo, said, "the way people used to live, near industry and walk to

Variable	Officials			Cities		
	#	%	n^*	#	%	n^*
Cities with regional government	—	—	—	2	10	21
Number that indicated city's form of government matters	17	71	24	12	67	18
Mean score attributed to city's relationship with county government †	7.65	—	17	7.42	—	12
Mean score attributed to city's relationship with state government †	7.76	—	25	7.59	—	16
Mean score attributed to City's relationship with federal government †	7.63	—	24	7.43	—	15
Number that indicated intergovernment relationships matter	19	79	24	12	75	16
Planner	17	47	36	—	—	—
Economic developer	10	28	36	—	—	—
Brownfields coordinator	6	17	36	—	—	—
Other	3	8	36	—	—	—
Mean years in position	5.86	—	36	—	—	—
Mean years in city	10.75	—	36	—	—	—
Planning undergraduate degree	5	14	36	—	—	—
Other undergraduate degree	7	19	36	—	—	—
Planning graduate degree	11	31	36	—	—	—
Other graduate degree	13	36	36	—	—	—

* number of respondents from sample

† Measured on a scale from 0 to 10 (10 being the best)

work, is not how people live anymore." Another official reiterated the point, saying that the historic land-use pattern of his northeastern city meant that "industry co-mingled with housing for workers." In Hartford, Connecticut, both Mark McGovern (Director of Economic Development) and Bruno Mazzula (Director of Housing and Property Management) agreed that the story was quite different there. Mr. McGovern said, "we don't have the classic neighborhoods, we have our industry segregated" and he could only think of six HI-TOADS within the city borders. Mr. Mazzula likewise thought that the city was not a big generator of HI-TOADS because even the most unsightly and dangerous abandoned property is far removed from the people whom it might impact. He estimated there to be only five HI-TOADS in Hartford. For cities without HI-TOADS, lack of industry was pointed to by interviewees.

Table 3.3

Active or Inactive HI-TOADS from 1995 to 2005

City	State	Number of former HI-TOADS	Number of current HI-TOADS
Baltimore	Maryland	3	4
Buffalo	New York	1	75
Camden	New Jersey	1	42
Cincinnati	Ohio	2	4
Dayton	Ohio	2	3
Gary	Indiana	2	5
Hartford	Connecticut	1	5
Knoxville	Tennessee	3	4
Louisville	Kentucky	3	6
Macon	Georgia	3	0
New Bedford	Massachusetts	1	4
New Haven	Connecticut	1	6
Newark	New Jersey	3	2
Pittsburgh	Pennsylvania	1	5
Richmond	Virginia	3	7
Rochester	New York	1	2
Savannah	Georgia	3	5
Scranton	Pennsylvania	1	4
Syracuse	New York	1	3
Trenton	New Jersey	1	6
Youngstown	Ohio	2	5

"I don't think we have anything under your HI-TOADS definition. . . . not a whole lot of industry in [my Southern city] . . . Your Rustbelt, Midwest cities, they have them."

Beyond historic land-use patterns, many respondents saw a poor real estate market as an important determinant of whether they had HI-TOADS. These cases where land was not valuable meant that sites that otherwise would have been reused were abandoned because nobody wanted them. Conversely, J. R. Capasso, Brownfield Coordinator for Trenton, New Jersey, remarked "now we have some market forces at work in Trenton, but it doesn't really help in those HI-TOADS areas."

However, in some cities, the market forces are strong enough that public support is not even needed. One official commented that "the market is so high" and it allows much adaptive reuse to be done by the private sector.

An official from a medium-sized southern city echoed this sentiment: "There's been tremendous brownfields development in the city in the last three or four years, but it's all been private, they don't tell us about their deals." It is in these strong market cities that HI-TOADS are either not a problem at all or are a minor one.

Several local officials also spoke about environmental regulation as a cause of HI-TOADS. In describing how environmental regulations changed in New Jersey, one official explained the evolution from laws that hindered to laws that help: "there was ECRA—Environmental Cleanup Responsibility Act, every time there was a sale of property, the site had to be cleaned up . . . it effectively shut down the sale of industrial urban properties. . . . Changed by ISRA [Industrial Site Recovery Act], brought in HDSRF [Hazardous Discharge Site Remediation Fund] and deed notifications and restrictions . . . provided for industrial-level cleanup." In New Jersey, the movement from ECRA to ISRA meant that contaminated properties could begin to be transferred again. Another New Jersey official critiqued ECRA "when DEP [New Jersey Department of Environmental Protection] came in with their standards [for property transfer] that is when the real problems started."

Several of the local officials described a particular type of HI-TOAD site, where industrial activity changed over time to a "bottom-feeder" level. According to an economic development official, "They're not very marketable for good, solid industrial uses [because of poor access and close proximity to residential areas], so low-end uses gravitate there, then those low-end uses bring down the neighborhood."[7] One example of this change was when a major manufacturer departed a building and a series of small-scale tire storage companies occupied the space. The quantity and quality of jobs was perceived by some officials as lower for these bottom-feeders. A city planning director further explained the problem: "The process is that they were not reoccupied to their fullest use [after the original owner left], there was no reason or benefit to reoccupy the higher stories [because higher floors do not typically meet modern industrial needs]. When they moved out the real blighting in there would occur." This "bottom-feeder" HI-TOAD site appears to have important economic and fiscal impacts, but its neighborhood impacts are unclear. But the reference to obsolete industrial configurations was a common theme throughout the interviews and is an essential problem that these local officials are concerned about.

What Problems do HI-TOADS Generate?

In trying to study the HI-TOADS phenomena, I sought to understand whether local officials viewed these sites as problems in their neighborhoods and in the city as a whole. Part of asking the question was to ascertain whether local officials believed that HI-TOADS caused a diminution of neighboring property values, whether they believed that a poor neighborhood caused HI-TOADS, or, as presented in chapter 2, whether the officials believed that both were true.

Twenty-four (73 percent) of the participating officials agreed that neighborhoods with HI-TOADS had lower property values than other neighborhoods in their city. One planner said "the further you get from a big, hulking, industrial, abandoned building, the nicer the neighborhood is." Another official, describing HI-TOADS neighborhoods in Buffalo, said "they struggle as far as property values . . . [only] the lowest stratification in the city" live in those neighborhoods. "Much lower, 40 to 60 percent lower" said Jeff Chagnot, Economic Development Director for the City of Youngstown, Ohio. Larry Stidt corroborated Mr. Chagnot's estimated value in describing the Delco and the Soccer Stadium sites in Rochester, New York: "property values are much lower in that area . . . upper $20,000s to $30,0000s [per housing unit] . . . versus $55,000 to $60,000 for housing in the rest of the city . . . housing there is much more modest."

Mr. Stidt touched on the issue of causality in his response: "Housing there is much more modest." Are the HI-TOADS neighborhoods less expensive than other neighborhoods because of the HI-TOADS or because of other confounding factors, such as the size, age, or type of the housing stock? In describing a neighborhood around a shuttered manufacturer, one midwestern official said: "It's an older section, it's within one-half mile of an operating steel mill, it's hard to say what effects the real estate values of the neighborhood. Values are probably lower." Many of the neighborhoods that host HI-TOADS have low property values compared to other city neighborhoods because of ongoing industrial operations. "Around here it's the operating sites that hurt value" one brownfield coordinator remarked. That is, an operating landfill or steel mill could be the reason that a neighborhood has low property values (while of concern to neighborhood well-being, such operating sites are Locally Unwanted Land Uses [LULUs], not

HI-TOADS; Popper 1981). According to some of the interviewees, once operations cease and the site becomes abandoned, neighborhood-wide property values may not be further dampened.

While there is some disagreement among officials about whether HI-
TOADS are only in neighborhoods with low property values, 28 of 30 respondents agreed that HI-TOADS represent a threat to their neighborhoods. "Yes,
I think they do, that's why we're focusing on them" responded Mr. Mazzula
of Hartford. "These places are treading water," said a northeastern planner.
But just as some officials felt that confounding factors beyond HI-TOADS
affected property values, some were skeptical of whether HI-TOADS were
the most important challenge to neighborhood stability. Mr. McGovern of
Hartford agreed that HI-TOADS are a threat, "but quite frankly, I don't
think those threats are any more than any other threats, for example vacant
storefronts." Others felt that historic land-use patterns played a more important structural role in neighborhood stability than any single HI-TOAD
site. "Once there was a slaughterhouse, those types of uses are never going
to be high value. It's still pretty scary down there . . . pretty disturbing," said
one planner.

While most interviewees agreed that HI-TOADS were a threat to neighborhood stability, a few mentioned specific problems: (1) a chilling effect
on investment, and (2) residential undesirability. Scott Alfonse, Brownfields Coordinator for the City of New Bedford, Massachusetts, expressed
his concern about how the reuse of an existing HI-TOAD site is in limbo
and the way that hurts investment activity in the neighborhood. "A lot of
people have put their activities on hold to see what will happen" in terms of
buying, selling, and renovations. Mr. Alfonse and other officials believe
that the fear of unknown development of unknown size and scale (or
worse, further dereliction) stalls investment activity in the neighborhood
around a HI-TOAD site.

A second particular problem related to neighborhood stability is the
question of residential desirability. The extent to which a neighborhood
features amenities that attract and retain residents is of particular concern to some officials when asked about neighborhood stability. Mr.
Cammarata was one who forcefully asserted a resident-centered notion of
HI-TOADS: "If the HI-TOADS were cleaned out and replaced with
something conducive to quality of life, it would be an improvement." A
brownfield coordinator in the Northeast, describing existing HI-TOADS

neighborhoods, was also focused on people and their housing choices. "Generally, I would say people would rather live elsewhere." These officials are emphasizing HI-TOADS' impact on people and their housing choices in contrast to loss of fiscal stability or out-migration of jobs, the more popular concerns of other local officials. This perspective of stability is important later in the following chapter, in the section on policies that cities have adopted to address HI-TOADS, a vast majority of which do not emphasize residential desirability.

I also asked the local officials whether HI-TOADS hurt surrounding areas in ways other than property values. As described in the above methodology section, not all questions were asked of all research participants. Only 22 of a total 38 interviewees provided answers to this question. There were two groups of non-respondents: the interviewees who were insistent during the earlier portion of the interview that HI-TOADS were not a problem in their city and the interviewees who could not emphasize enough the devastation caused by HI-TOADS. For the first group, I did not ask them about impacts because they made it clear that they did not feel that HI-TOADS were of concern and in many cases were standoffish. For example, one southern planner interrupted me and exclaimed, "single-family vacant buildings are the problem!" In an effort to keep the interview moving along in a socially graceful manner, consistent with Kvale's (1996) view of interview research as a conversation, I skipped the question. For the other group, the respondents already had indicated that they thought that HI-TOADS were devastating to their communities.

Of the 22 officials who answered the open-ended question about the kind of impacts that HI-TOADS generate beyond diminished property values, 16 (72.7 percent) mentioned crime, 10 (45.4 percent) mentioned dumping, 6 (27.2 percent) mentioned arson, and the following other impacts were mentioned at least once: safety or public health, litter, aesthetics, lost opportunities, brownlining, and lost confidence. [8] It is worth noting that those other impacts can cause a decrease in property values and are not likely independent indicators of HI-TOADS.

HI-TOADS can serve as a haven for criminal activity. As one official put it, "Any abandoned dark place is a good place for criminals and troublemakers to hang out." Local officials reported that robberies, drugs, prostitution, and illegal salvaging emanate from HI-TOADS. While seemingly a minor offense, Mr. Chagnot of Youngstown complained, "we've had people arrested who were looking for scrap metal . . . It draws police resources from neighborhoods because they are busy watching these properties."

Arson represents a unique threat to neighborhoods surrounding HI-TOADS. Suprisingly, only six officials responded that HI-TOADS are to blame for arson, but those who did were quite concerned. Mr. Stidt recounted the story of a recent fire at the Delco Plant in Rochester where an act of arson spread to nearby residential property and caused extensive damage. And in New Bedford, Mr. Alfonse linked a HI-TOAD site with a neighborhood mindset: "Certainly has had a psychological impact on the neighborhood . . . the fire chief was once quoted as saying 'if those buildings caught fire, we could have to evacuate homes in a 12-block radius,' that really got to people." Partially in response to the neighborhood hysteria generated by the chief's comments, the city paid to demolish those buildings so that today the site remains vacant, unimproved land. Neither of the two New Bedford officials interviewed thought that the site presently met the definitional criteria to be a HI-TOAD site.

The definition used for HI-TOADS includes properties that are at least partially abandoned. Some HI-TOADS discussed in the interviews had some on-going operations at the site, which in most cases also involved some level of protection and maintenance of the site. For cases where the HI-TOAD site was fully abandoned, dumping appeared to be a more serious concern. "A lot of illegal dumping goes on in these sites, it's hard to keep people from getting in," remarked Mr. Chagnot of Youngstown. One brownfield coordinator emphasized the correlation between dumping at HI-TOADS and property values in that neighborhood, "anytime you have a lower-value neighborhood and abandoned buildings, you get fly dumping—when people come in with pick-up trucks and dump—55-gallon drums of bad stuff." But in one coastal city, the dangers of dumping become exacerbated. "Many of our sites are coastal, it's a real problem for us." When illegal dumping occurs on these coastal sites, the waste can simply float away and cause widespread environmental problems.

The officials largely believed that these HI-TOADS in their cities represented a problem. Beyond property values, crime and dumping were the most frequently mentioned impacts, as well as a contribution to neighborhood malaise. Camden, in many ways one of the epicenters of HI-TOADS, suffers from a complex, multifaceted set of social, economic, and fiscal problems, of which abandoned properties are only a part. But local officials there were insistent that HI-TOADS represented a critical challenge to their city. Likewise, in cities with only one or two HI-TOADS, the local officials expressed their concerns about the ways that the HI-TOADS drag down their neighborhoods. Of the 38 officials interviewed, only a handful did not see HI-TOADS as a serious problem.

[4]

A National Perspective on What Cities Do about HI-TOADS

This chapter examines what cities do to address HI-TOADS and how well do they do it. The unit of analysis here is the city. In this section, I will present the quantitative interview results for both cities (n=21) and officials (n=38) for comparative purposes. I asked each official several questions to probe the role that his or her city plays in managing HI-TOADS and the specific policy tools used by the city. To be clear, I asked about cities' existing policy tools and those that are developed with the express purpose of addressing HI-TOADS. How do existing city policy tools (such as zoning, tax increment financing, abandonment taxes) and HI-TOADS–specific policy tools (such as a neighborhood development plan or an interim use of a HI-TOAD site) work to address HI-TOADS? What emerged from the interviews was a set of fourteen distinct policy tools (see table 4.1).

Chapter 2 introduced three types of policies used to organize much of this study: economic development, community empowerment, and healthy environment policies. Despite my ability to break down fourteen policies into three categories, the respondents generally did not break them down. Rather, they used multiple policies, sometime mixing two or three policies together in a single project or plan. In this section of the chapter, I present the policies separately and then discuss their combined use.

In a series of three questions, I asked each local official about the role that their city plays in addressing HI-TOADS and what kinds of policies their city has adopted. If the official was not able to respond to these questions, I asked about specific policies such as development incentives or zoning.

Table 4.1
Policies Adopted to Address HI-TOADS

Policy set	Policy tool	Number of cities	Percent of cities	Number of officials reporting	Percent of officials reporting
Economic development	Planned development districts	11	55	11	31
Healthy environment	Mixed-use	4	20	5	14
Healthy environment	Form-based zoning	2	10	2	6
Economic development	Condemnation	11	55	20	56
Economic development	Ease development review/approval	7	35	11	31
Economic development	Tax abatements/grants/loans	17	85	29	81
Economic development	Empowerment zone	9	45	12	33
Economic development	Shovel-ready	16	80	27	75
Economic development	Assume environmental liabilities	8	40	10	28
Healthy environment	Adaptive reuse	7	35	9	25
Healthy environment	Interim use	3	15	3	8
Community empowerment	Neighborhood planning	7	35	11	31
Healthy environment	Asset-based planning	1	5	1	3
Healthy environment	Brownfields office	7	35	11	31
		n=20		n=36	

Policy Tools for Addressing HI-TOADS

Zoning

Zoning is one of the most powerful tools that cities have to influence the built environment (Platt 1996). But its most potent form comes during a city's initial growth. Using zoning effectively to address redevelopment is a real challenge for today's cities. Zoning can be cumbersome, difficult to revise, and in many cases difficult to interpret. Zoning is also easy to manipulate: obtaining variances or otherwise changing code on an ad hoc basis is common practice. Many officials referred to their zoning as "outdated" and described it as inconsistent with current master planning. Officials in a few cities reported that they recently had updated their zoning. "Our zoning is pretty progressive in terms of not hampering development," said Ms. Golumb, referring to a 1998 update to Pittsburgh's zoning code.

Zoning can fall into either an economic development set of policies or a healthy environment set of policies. When the zoning is intended to promote new growth and development, then it can be considered a tool of economic development. When zoning is used primarily to protect public health and the environment, then it is a tool of healthy environment policies.

The three zoning-related tools mentioned in the interviews are Planned Development Districts, mixed-use zoning, and form-based zoning. In total, 8 of the 20 cities use one of these zoning tools in addressing HI-TOADS.[1] Planned Development Districts (PDDs)are the most popular zoning tool, utilized by 7 cities (35 percent). These PDDs are enacted into cities' zoning codes, in most cases as overlays. One common way that PDDs are codified as through a state government-sanctioned redevelopment plan. Once adopted, a redevelopment plan establishes certain zones in a redevelopment area and provides for a high level of control of uses, densities, and bulk. New Jersey, Connecticut, and New York officials were particularly vocal about using PDDs through their state's redevelopment planning process. According to Mr. Camarrata, New York State's urban renewal process allows Buffalo to better control for the kinds of uses that will occur at a development site than typical zoning would. They have used it to successfully and "kept out certain uses, like chemical companies." New Haven has used a coastal management overlay district to address abandonment of industrial properties along the water. Hartford adopted Industrial Reuse Ordinance Floating Districts (I-RODS) to better facilitate residential conversion of industrial properties in several sections of the city. Cities have adopted these PDDs in order to provide for greater developer flexibility in order to encourage private-sector reuse of former industrial properties, while controlling for densities, bulk, and uses in a more flexible manner than the regular zoning code would allow.

Many of the PDDs referred to above include elements of mixed-use zoning. The Euclidean zoning model is based on the premise that separation of industrial and residential uses is of paramount importance. Many planners and their agencies have moved away from this thinking and have adopted mixed-use zoning to allow for industrial and residential uses to be in very close proximity. In most cases, strong provisions are integrated into these mixed-use ordinances to provide protection for the residents against noxious odors or other pollution. However, one official expressed strong concern about this practice: "Movement to gentrification is the biggest risk to industrial [development . . . these industrial sites are] not an appropriate place for residential." The public health risks may be too great for large residential populations, which may account for the fact that only 4 of the 20 (20 percent) cities use mixed-use zoning. Other officials fear that an influx of residents into industrial zones could limit available space in their city for industrial development. "This is our prime industrial land" said one midwestern planner. Ms. Golumb explained that the City of Pittsburgh has hired an outside consultant to answer the question "How much of the land in the city is important to keep as industrial land?"

Officials in Louisville, Kentucky, reported that their city uses form-based zoning as a tool to address HI-TOADS. This relatively new zoning approach was developed by New Urbanists as a way to impose more cohesive urban design through a higher level of regulation of the design and form of new development (Langdon 2005b). With less of an emphasis on setback requirements, floor-area ratios, and building heights than conventional zoning, form-based zoning instead prescribes urban design aesthetics with the intention of creating "places whose proportions generate a pleasing sense of enclosure" (Langdon 2005b, 27). Like mixed-use, form-based zoning may provide developers with a greater degree of autonomy and flexibility to redevelop a HI-TOAD site. This view of conventional

HI-TOAD Site Profile: Colt Fiream Plant, Hartford, Connecticut

Formerly a major manufacturer of firearms off Interstate Highway 91, Colt's complex of twelve buildings on about 20 acres was partially abandoned at the turn of the millennium. Through a $110-million, public-private partnership and in concert with a Planned Development District zoning overlay, the historic site will be reused adaptively for "a 24/7 live-work complex—housing, offices" and a park, according to Mr. McGovern. The large-scale size of the property, its historic status, and environmental challenges made for a difficult redevelopment. However, with a combination of federal and state money, tax–increment financing through the City of Hartford, and the developer bringing in Historic Tax Credits, the Planned Development District provides an additional in-kind support for the project from the city.

zoning as a barrier to redevelopment was not universally expressed in the interviews. Rather, most officials saw little connection between how a HI-TOAD site was zoned and its prospects for redevelopment.

Condemnation

The next policy mentioned by local officials as part of their efforts to address HI-TOADS is eminent domain or condemnation authority. Condemnation is used almost exclusively to promote economic development. In rare cases, it is used by cities to acquire dangerous and derelict properties so that they can be demolished or remediated.

As discussed in chapter 2, the use of condemnation as part of an urban redevelopment project is under tremendous fire since the 2005 Supreme Court ruling in *Kelo v. City of New London.* Despite the rash of state legislation, Congressional resolutions, and public outcry, over half of cities surveyed use eminent domain in addressing HI-TOADS. Of the 38 officials surveyed, 20 (52.6 percent) mentioned condemnation as a tool their cities employ when faced with HI-TOADS.

"[My] city has powers of condemnation and linked funding, if you can demonstrate potential for job growth" said a planner from a large Northeast city. In many states, if a city can demonstrate the potential for employment generation, then it wields strong condemnation powers. But

just because a city technically can act, that doesn't mean that it is politically feasible to do so. The controversy in New London stemmed partially
from the fact that the city sought to acquire homes owned and occupied
by middle-income residents. In the cases studied in this research, the cities
are seeking to acquire abandoned or partially abandoned industrial structures or vacant land. Nevertheless, many officials were sensitive to the political concerns of eminent domain.

Economic Development

The third class of policies are both economic development tools and part of
the larger economic development rubric identified in chapter 2. They include (1) easing of development review/approval; (2) tax abatements, grants,
and loans; and (3) empowerment zones. All but one official interviewed

HI-TOAD Site Profile: Frigidaire Site, Dayton, Ohio

Built in the 1920s as one of Fridigaire's primary manufacturing sites, the
now-abandoned buildings occupy a whole city block and have been targeted for redevelopment. Using the powers of eminent domain, the City
of Dayton, through its CityWide Development Corporation, has acquired
the site and is preparing it for development as part of its TechTown neighborhood redevelopment project. Dayton's Planning Director, John
Gower, spoke about how acquiring the Frigidaire site was the first in a
three-step process of getting the HI-TOAD site into a new use: "got property, did environmental work, did demo." Dayton's high crime and unemployment levels make a project like this politically feasible. However,
not all cities can move so swiftly in taking property for economic development purposes, as recent reactions to the *Kelo* decision have illustrated.
For example, one Connecticut planner expressed concern about his city's
ability to continue to acquire property through condemnation: "That entire program in New London was based on the Municipal Development
Program. We are worried because we have used that program successfully." The State of Connecticut's Municipal Development Program was
challenged in the *Kelo* case for permitting cities to acquire property
through condemnation to further economic development objectives.

reported using at least one of the three economic development tools in efforts to address HI-TOADS.

Eleven officials (28.9 percent) in seven cities (33.3 percent) discussed how they proactively ease development review or approval in an effort to better facilitate private reuse of HI-TOADS. "The city's role has been to grease the way for the private sector," stated one southern economic development official. Another southerner explained that the city's role is "navigating people through the re-development process . . . typically . . . many of these sites have industrial zoning . . . [that requires] special permits and re-zoning . . . It's about working people through this process and negotiating agreements to work through it."

By far the most frequently mentioned policy in the interviews was the use of tax abatements, grants, and loans to encourage or support HI-TOADS reuse projects (29 of 38 officials [76.3 percent] representing 17 of 20 cities [85.0 percent]). Officials viewed this as the most cost-effective manner to manage multiple sites, without the city exposing itself directly to high levels of liability. One brownfields coordinator explained her city's role: "Sometimes a business comes here and says they want to relocate . . . then we help them with getting set up [site preparation, environmental assessment, clean-ups, etc.]. The city has a really good incentive package . . . we can do a good job to help a business find a location." In summing up the array of loans, tax credits, and grants available to encourage redevelopment of HI-TOADS, one brownfields coordinator remarked, "we'll do whatever it takes to bring in redevelopment to the site. A pretty good bag of tricks."

In one New Jersey city, an economic development official explained that the city attempts to shepherd the multitude of funding opportunities to the benefit of redeveloping HI-TOADS (among other city priorities). "The city's role has been to try to coordinate public resources that are trying to address these properties." Many of the tax abatements, grants, and loans available to HI-TOADS redevelopers are either state or federally funded, so the city's role in this sense is to coordinate these resources in a strategic manner.

The third set of economic development policies identified through the interviews relied heavily on geographically defined zones and their accompanying financial benefits. Officials mentioned both state and federally designated zones: Empowerment Zones, Enterprise Zones, Empire Zones, and Brownfields Development Areas (New Jersey only). For those 12 officials (31.6 percent) in nine cities (42.9 percent), the geographically distinct boundaries of these zones provided an additional layer of power and allow pooling of resources to address HI-TOADS. In Rochester, Mr. Stidt explained that "most of these older facilities are in Empire Zones; we've applied for Commercial Urban Exemption program for some of these properties." In New York State, a Residential-Commercial Urban Exemption represents a ten-year graduated tax abatement for that portion of a building converted to residential uses. The exemption is just one of scores of unique powers bestowed upon municipal governments that have these state or federal zones covering their territory.

New Jersey's Brownfield Development Area (BDA) Initiative works in a similar way, focusing on an area-approach to brownfields redevelopment. As of November 2005, fifteen BDAs have been designated throughout the state (New Jersey Department of Environmental Protection 2006a). In these areas, HI-TOADS can be addressed in close coordination with other environmental investigation, remediation, and redevelopment at surrounding brownfields sites and the state provides additional attention and resources to the projects. Not to be outdone by its neighbor to the south, New York State also has a zonal incentive program called the Brownfields Opportunity Area (BOA) Program. Two cities in this study received BOA designation early in 2005, Buffalo and Syracuse. Like the New Jersey program, BOA-designated areas receive technical assistance grants and have priority consideration for other kinds of state funding.

The economic development programs discussed here drew heavily on neighborhood life-cycle theory, as introduced in chapter 2. In describing her city's efforts, Ms. Sundrla of Savannah said: "I think we're going to see real changes. The city has targeted 18 neighborhoods for redevelopment. . . . I think we'll see the majority of these blighted neighborhoods on their way to recovery." Ms. Sundrla's use of the phrase "on their way to recovery" is suggestive of a patient after surgery and reflects an acceptance of neighborhood life-cycle theory and its proposition that neighborhoods are analogous to living organism. Many of the officials saw economic development policies, in particular, as prescriptions for ill neighborhoods, reflecting a belief that a city's policy intervention will cure the ills of a neighborhood and help them to recover. This expressed ideology is quite different from the alternative neighborhood-change theory also introduced in chapter 2.

HI-TOAD Site Profile: South Buffalo Industrial Area, Buffalo, New York

A 1,200-acre conglomeration of former pig-iron and steel plants in South Buffalo represents one of the greatest HI-TOADS challenges faced by any city in this study. Buffalo already has difficulties to contend with: high crime, high unemployment, widespread housing abandonment, and depopulation. Planning for successor land uses to vast stretches of industrial land is not easy. However, New York State is trying to help through is pre-existing Empire Zones designation (which covers the industrial area) and more recently through its designation of the industrial area as a Brownfields Opportunity Area. The designation is helping the South Buffalo Industrial Area "move from a twentieth-century to a more twenty-first-century use, from an industrial sector to a service-oriented, commercial sector," according to one official. Through its use of economic development policies such as grants, tax abatements, and loans, the City of Buffalo and the State of New York are focusing their resources toward HI-TOADS in this unique and quite large zone.

Site-Work Policies

The first three classes of policy responses dealt with the ways that cities zone for, acquire, and promote the private reuse of HI-TOADS. This section describes the ways that cities implement policies to directly change the physical conditions of a property. Site-work activities are tools that fall under both economic development and healthy environment policies. Like zoning, site-work tools can be used to accomplish varying goals, depending upon how they are executed. In some cases, a city can accomplish multiple goals with a site-work project.

With the exception of just two cities, Louisville, Kentucky, and Rochester, New York, officials in all of the other cities in this study specifically mentioned site work as part of their role in addressing HI-TOADS. Local officials described four site-work-related policies: (1) shovel-ready, (2) assumption of environmental liabilities, (3) adaptive reuse, and (4) interim use.

Second only to tax abatements, grants, and loans, the shovel-ready policy was mentioned by 27 local officials (71.1 percent) in 16 cities (76.2 percent).

Consisting of a series of activities, including demolition, environmental site assessments, installation of infrastructure, and marketing, local officials rely heavily on shovel-ready policies to address HI-TOADS. In speaking about Dayton, Ohio's approach, Planning Director John Gower remarked: "One foundation of our broader economic development strategy [is that] in order to service development opportunities, we need development-ready land."

A northeastern planner embraced a toad metaphor in explaining the way that his city uses a shovel-ready policy: "We recognize that for these sites, we may need to do additional work . . . we try to do as best a makeover as possible, remove all the warts. Essentially minimize the risk of the investment . . . how could you promote and market these sites, you identify the most appropriate reuse and push it to prospective developers."

The popularity of this heavy-handed, proactive role for the city is somewhat surprising given an era of limited public resources and a retrenchment of the role of government in the marketplace. But based upon just such a shovel-ready policy, billions of local, state, and federal dollars have been invested at HI-TOADS included in this study: for demolishing factories (local, state, and EPA funding), conducting Phase I and Phase II environmental site assessments (local, state, and EPA funding), installing of new infrastructure (local, state, HUD, and U.S. Economic Development Administration funding), and marketing these remediated sites (local funding).[2]

Very much related to shovel-ready policies is the question of environmental liability. Over the last few decades, sweeping changes have occurred in the extent to which local governments are liable for environmental remediation through their redevelopment activities. Prior to CERCLA (the Comprehensive Environmental Response, Compensation, and Liability Act of 1980), local governments could be very involved in investigating, clearing, and owning contaminated property with little fear of liability. But after CERCLA and a string of state-level equivalent laws, local governments began to be implicated as potentially responsible parties even with the slightest level of involvement. But in recent years, with the ascent of brownfields laws and policies, local governments are once again willing to take on liability or are being shielded by new state laws (as in Indiana).[3]

Youngstown, Ohio's Deputy Director of Planning, William D'Avignon spoke most definitely on the matter saying, "we've taken a hold-harmless position for environmental issues." In an effort to attract new development onto the city's HI-TOADS, the city is telling developers that it will share in the liability with the new developer if environmental issues arise at a site. Ten officials (26.3 percent) in eight cities (38.1 percent) reported using an Assumption of Environmental Liabilities policy to help redevelop HI-TOADS. Youngstown's position was the extreme; most of the other cities are silent on the question of future contamination in their dealings with developers, but are involved enough in deals that they are exposing themselves to liability knowingly. One official explained that her city had not yet adopted such a policy, but expected that it will one day. When that day comes, she points out there are tools to keep the city protected on some level, "third-party liabilities, cost-cap insurance, that's what people are going to . . . we haven't used them yet, but we will . . . everybody's protected. That's opened up doors, I think."

While shovel-ready policies are concerned with getting a property cleared and ready for a developer to build, its antithesis is adaptive reuse. Adaptive reuse policies seek to preserve existing structures in order to foster greater resource efficiency, to protect historically significant buildings, or to maintain an existing urban or neighborhood fabric (Burchell and Listokin 1981). Nine officials (23.7 percent) in seven cities (33.3 percent) reported the adoption of adaptive reuse policies in addressing HI-TOADS. One southern official described his city government: "We are a preservation force to be reckoned with." He further went on to characterize local historic architecture by saying "it's the best thing going for the city."

Many officials spoke about the role that adaptive reuse can play in enhancing tourism in their cities. For example, at the Drayton Street Gas Station in Savannah, described on page 37, the "city is preventing the gas station from being demolished . . . [it is the] city's role to protect those structures." By protecting the historic gas station structure, Savannah officials feel that they can keep the city's downtown historic district intact and advance its tourism draw.

Another southern official described his office's approach to HI-TOADS:

> The [city's] role has been to encourage sponsoring National Register designation to facilitate adaptive reuse of buildings . . . [we] more than doubled the number of buildings listed in the city—the majority of buildings eligible for listing are listed. We have been leading the country with the use of the Historic Preservation Tax Credit Program.

Among the site-work class of policies, the final one is interim use. Where the environmental investigation, demolition, marketing, and redevelopment of sites can take years if not decades, three officials (7.9 percent) in three cit-ies (14.3 percent) mentioned the interim uses of sites as a policy response to HI-TOADS. Here, the officials are concerned with mitigating against the ongoing, neighborhood-wide impacts of HI-TOADS by making a new, al-beit temporary, use of the site. In one New Jersey city, "sometimes we can use [the HI-TOAD site] for a parking lot, for an interim use . . . some we used for circuses." Mr. Alfonse described the interim use of a portion of the Morris Cutting Tools Plant in New Bedford, Massachusetts: "It's a two-block parcel, most of the contamination is on only one block . . . on the cleaner of the two, we took down the fence, loamed and seeded it, and it's a green space . . . people in the neighborhood use it, usually for festivals." Whether for circuses, parks, festivals, or parking lots, these interim uses have the potential to mitigate the impacts of HI-TOADS.

Planning

Lastly, I present the fifth category of policies mentioned by local officials: planning. Chapter 2 defined planning and described the differences between government planning and the activity of planning, namely codifi-cation being a hallmark of government planning. Here, the focus is on local officials' planning activities. Planning is predominately a tool for either community empowerment or healthy environment policies, while in some cases economic development objectives play an important role. From the interviews, three planning activities emerged: neighborhood planning, asset-based planning, and brownfields offices. It is also worth reiterating

---

HI-TOAD Site Profile: Tobacco Row, Richmond, Virginia

Stretching over a mile, Tobacco Row (as its namesake implies) was for hundreds of years a string of tobacco warehouses and storage sheds. That began to change in the last couple decades of the twentieth century when, through an aggressive city effort to promote adaptive reuse, one by one the buildings received federal National Register designation and subse-quently were renovated to their original appearance. The eighteenth- and nineteenth-century structures were converted into offices, shops, and condominiums by a variety of different developers. "Tobacco Row is es-sentially breathtaking . . . breathtaking urban space" said one local official.

that any professional engaged in planning is a planner, no matter what office or agency she works in.

Twelve local officials (31.6 percent) mentioned one or more planning activities as part of their response to HI-TOADS. Considering the severity of HI-TOADS and the capacity these cities have in terms of professional planning, it was surprising to find that a majority of cities do not see planning as an important part of their response to HI-TOADS.

Virtually all cities do neighborhood planning. But only seven cities (33.3 percent) do it as part of a strategy to address HI-TOADS. Only 11 officials (28.9 percent) mentioned neighborhood planning as a strategy, while almost all officials focused on site work. This focus on site over neighborhood represents a real problem for addressing what is by definition a neighborhood problem. Mr. Stidt described the use of neighborhood planning in Rochester, New York: "Neighborhood planning studies were used to identify, from a neighborhood perspective, what [residents] would like to see there, and then to use that to secure funding. An overall, area-wide plan . . . In the stadium site, there was a second round of funding which included $1,000,000 in neighborhood reinvestments."

In Rochester and other cities, local officials tried to empower residents in neighborhoods affected by HI-TOADS to help determine reuse visions. This pseudo-grassroots planning process provides for neighborhood impacts to be considered paramount to city-wide or even state-wide concerns.[4]

While Dayton's Mr. Gower was the only official to mention it, asset-based planning is a unique approach to addressing HI-TOADS. He explained, "we have to plan for a shrinking city . . . Our department is more focused on an asset-based approach to major institutions: education and health care." When faced with the regular disappearance of major industries and institutions, Dayton is focusing its policy agenda on asset-based planning to ensure that its "growing and thriving businesses and institutions" receive the greatest attention. Rather than fighting to keep dying industries afloat, Mr. Gower insists that the city's limited resources should go toward supporting those that are strong. This planning approach focuses the city's attention for HI-TOADS reuse toward outcomes that strategically support select "businesses and institutions."

Mr. Gower's discussion of asset-based planning appeared to draw on the alternative neighborhood-change theories introduced in chapter 2. While other officials emphasized the need to arrest the deterioration of city neighborhoods or the need to bring back once-thriving districts, asset-based planning is an exception to this discourse. Under asset-based planning, the city government is not making a claim that given neighborhoods are dying or dead; rather, they are identifying the city's key assets and investing in them. Borrowing from the ideas of alternative neighborhood-change theorists, asset-based planning escapes the grand narrative of urban decline and seeks ways to enhance local residents' quality of life.

Eleven officials (28.9 percent) in seven (33.3 percent) cities spoke about the importance of their brownfields offices in addressing HI-TOADS. A relatively new phenomenon, scores of cities began to establish brownfields offices within their city governments as the EPA launched its Brownfields Pilot Program in the late 1990s. While federal funding continues to support many brownfields offices, officials reported that local and state funding now pays the lion's share of the bill. These offices are most often a city's central point of contact for HI-TOADS reuse projects, serving to coordinate the city's departments and other resources. Mr. Capasso of Trenton explained that the "city has established a Brownfields Program to address environmental issues and management grant/revolving loan funds" as one of its main tools to address HI-TOADS. Cincinnati's Strategic Program for Urban Redevelopment functions as a brownfields office and is staffed by interdisciplinary teams that work around one of thirteen districts in the city, and most importantly, "it's funded." Mr. McGovern lamented the lack of a brownfields office in his city: "We don't have a dedicated brownfields program in the city of Hartford [Connecticut]. We had someone, but it was eliminated four years ago."

HI-TOAD Site Profile: Oak Street Junkyards, Rochester, New York

The 10-acre Oak Street Junkyards were a long-term LULU in a residential neighborhood northwest of Rochester, New York's downtown. After the junkyards were abandoned, the city acted by initiating a neighborhood planning effort. Through a series of public meetings and workshops, city officials learned that neighborhood residents wanted to see a soccer stadium built on the site. Using that local knowledge, the city developed a neighborhood plan exploring the junkyards site and its context. Through that plan, the city has attracted millions of dollars in external grants and loans both for the site itself and for the neighborhood.

How Cities Use Policy Tools

Of the policies described above, few were used in isolation. Rather, they were combined in various ways to address the HI-TOADS problem from multiple perspectives. Figure 4.1 shows the intersections between each of the five groups of policies. Each cell shows how many cities used both groups of policies in addressing HI-TOADS. For example, the cell at the intersection between planning and zoning policies indicates that four cities used both categories of policies in addressing HI-TOADS. The mean cell has a value of 8.5 cities, with the highest cell value of 17 (combined use of economic development and site work). Among the 20 cities sampled, figure 4.1 suggests that combining of policies is quite common.

Table 4.2 presents the sum of policy tools adopted for each city. Cities' HI-TOADS policies are spread throughout the five categories, with a low of only two policies (Knoxville) and a high of eight policies (Richmond and Pittsburgh). The average city adopted 5.25 policies. The average city also chose from a broad array of policies, not only from a few in the same group of policies. On average, cities drew from 3.3 of the 5 policy groups (66 percent). New Haven was the only city that used policies from each of the five groups. When I asked Mr. Luongo of New Bedford which of his city's policies have worked in addressing HI-TOADS, he responded "what's worked is a combination," citing the city's good relationship with environmental regulators, its aggressive record acquiring properties, and its reuse-planning efforts.

Bivariate cross-tab analysis revealed some interesting patterns of combination. In table 4.3, I present a cross-tab of planning and site-work policies. Cities that employed two planning policies accounted for only four of the 18 cities (22 percent) that used at least one site-work policy. On the flip side, those cities that did not employ any planning policies accounted for nearly half of the cities that used at least one site-work policy. Together,

	Zoning			
Condemnation	5	Condemnation		
Economic development	8	11	Economic development	
Site work	6	9	17	Site work
Planning	4	5	10	10

n = 20

FIGURE 4.1. Combined use of policy tools to address HI-TOADS.

Table 4.2
Sum of Policies Adopted for Each City

City	State	Zoning	Condemnation	Economic development	Site work	Planning	Total number of policies
Baltimore	Maryland	0	1	1	2	1	5
Buffalo	New York	0	0	3	2	2	7
Camden	New Jersey	0	1	1	1	1	4
Cincinnati	Ohio	0	0	2	1	2	5
Dayton	Ohio	0	0	0	2	1	3
Gary	Indiana	0	1	2	2	0	5
Hartford	Connecticut	1	1	1	3	0	6
Knoxville	Tennessee	0	0	1	1	0	2
Louisville	Kentucky	2	1	1	0	0	4
New Bedford	Massachusetts	1	0	1	3	2	7
New Haven	Connecticut	1	1	2	2	1	7
Newark	New Jersey	0	0	3	1	0	4
Pittsburgh	Pennsylvania	2	0	2	4	0	8
Richmond	Virginia	2	0	3	2	1	8
Rochester	New York	1	1	2	0	1	5
Savannah	Georgia	2	1	1	1	0	5
Scranton	Pennsylvania	0	1	1	2	0	4
Syracuse	New York	0	0	2	1	2	5
Trenton	New Jersey	0	1	3	2	1	7
Youngstown	Ohio	0	1	1	2	0	4

Table 4.3

Cross-Tab between Planning and Site Work

		Number of planning tools			
		0	*1*	*2*	*Total*
Number of site-work tools					
0	Actual count	1	1	0	2
	Expected count	0.9	0.7	0.4	2
	Percent within site work	50	50	0	100
1	Actual count	3	1	2	6
	Expected count	2.7	2.1	1.2	6
	Percent within site work	50	17	33	100
2	Actual count	3	5	1	9
	Expected count	4.1	3.2	1.8	9
	Percent within site work	33	56	11	100
3	Actual count	1	0	1	2
	Expected count	0.9	0.7	0.4	2
	Percent within site work	50	0	50	100
4	Actual count	1	0	0	1
	Expected count	0.5	0.4	0.2	1
	Percent within site work	100	0	0	100
Total					
	Actual count	9	7	4	20
	Expected count	9	7	4	20
	Percent within site work	45	35	20	100

these results suggest that planning and site work may be inversely related. Alternatively, the results could be interpreted to mean that planning and site work are substitutes for each other.

A similar pattern emerges from a close examination of table 4.4, which shows the cross-tabs between planning policies and condemnation. For those cities that employed two planning policies, none of them used condemnation. And for those cities that did not employ any planning policies, two-thirds did use condemnation. While site work and planning feasibly are inversely related, it is unlikely that a true inverse relationship exists between planning and condemnation. Many state laws require that cities develop comprehensive plans prior to condemnation for redevelopment purposes. The results in table 4.4 likely reflect a relationship between those cities actively engaged in planning and working closely with communities

Table 4.4

Cross-Tab between Planning and Condemnation

		Number of planning tools			
		0	*1*	*2*	*Total*
Number of condemnation tools					
0	Actual Count	3	2	4	9
	Expected count	4.1	3.2	1.8	9
	Percent within condemnation	33	22	44	100
1	Actual count	6	5	0	11
	Expected count	5	3.9	2.2	11
	Percent within condemnation	55	45	0	100
Total					
Actual count		9	7	4	20
Expected count		9	7	4	20
Percent within condemnation		45	35	20	100

and those that are skittish about the recent backlash to the abuse of eminent domain. That is, cities may stress that they focus on planning and not condemnation to address HI-TOADS, but in reality do both.

Local officials utilize a diverse range of tools to address HI-TOADS. With few exceptions, most of those tools do not impose requirements as to what the new use of the property will be. Instead, a combination of local officials, private developers, and citizen activists tend to dictate the ultimate reuse of HI-TOADS. In this final results section pertaining to my questions regarding what policies cities use to address HI-TOADS and how successful they are, I begin by describing the range of reuses envisioned, planned for, and acted upon at HI-TOADS included in this study. Then, I present an assessment of the outcomes of those projects and overall assessments of each of the 14 HI-TOADS policies.

Table 4.5 features a list of the land uses envisioned, explored, planned, or acted upon at current or former HI-TOADS.[5] Because nearly all of the HI-TOADS included in this study are former industrial properties, and they are almost all legally within industrial zones, industrial reuse would be legally feasible at most HI-TOADS. Whether industrial reuse is financially feasible hinges mainly on the extent of a building's obsolescence. In describing the abandonment of industrial structures in Trenton, one official said, "there was no reason or benefit to reoccupy the higher stories" because higher floors typically do not meet modern industrial needs. "None of the multi-floor industrial buildings have included any new industrial, except maybe at Colt," Mr. McGovern added.

But as Mr. Chagnot pointed out, just because a city has a heavy industrial past, that does not mean it seeks to have a heavy industrial future. In Youngstown, he is working to "either redevelop them as greenspace or a *light* industrial usage" (for example, warehousing or assembly plants). While cities all seem to consider industrial (especially light industrial) uses, they also pursue a wide range of other uses.

Excluding industrial uses, office, retail, open space, and residential account for the four most-often cited uses by local officials. At the Delco site in Rochester, "Neighbors had expressed an interest in it being put back into residential use. Salvation Army has proposed a community center for the site. They're looking at anything that shores up quality of the neighborhood, that reinforces the residential base." Here Mr. Stidt explained that whatever the reuse may be, the city government and neighbors are concerned primarily with achieving an improvement in the "quality of the neighborhood." This neighborhood-impact-oriented reuse planning was not echoed by many other officials. Most were more focused on job creation, tax ratables, and economic multiplier effects.

Table 4.5

Non-Industrial Land Uses Explored, Envisioned, Planned, or Acted Upon at Former or Current HI-TOADS

Land use	Number of cities	Percent of cities	Number of officials	Percent of officials
Office	13	61.9	13	34.2
Retail	11	52.4	13	34.2
Open space	10	47.6	17	44.7
Residential	10	47.6	14	36.8
Educational	3	14.3	3	7.9
Tourism	2	9.5	5	13.2
Entertainment	2	9.5	4	10.5
Hotel	1	4.8	1	2.6
	n=21		n=38	

Measured in terms of elimination of HI-TOADS (through their reuse or redevelopment), 24 of 35 officials (68.6 percent) felt that their city's policies were successful (four thought they were not successful and seven did not know). Jerry Detore, Executive Director of the Urban Redevelopment Authority of Pittsburgh, said "I think we've been very successful; the major brownfields sites we've developed are all high-quality, positive tax cash flow for the city, attracting suburban life into the city." This response stands in stark contrast to what Mr. Stidt was describing above in terms of improvements in neighborhood quality of life.

Because HI-TOADS by definition indicate depressed property values, I asked the officials whether property values have risen to the extent that the HI-TOADS neighborhoods reach are no longer affordable for middle-class residents or small businesses. Here, I am trying to understand whether HI-TOADS reuse efforts are spurring gentrification. Only 12 of 35 officials (34.3 percent) agreed that they had, with one saying he did not know. An official in a southern city was one who said yes. She explained, "That has been a big concern. We developed a gentrification policy for the city to address not only affordable housing but affordable commercial space, integrated into our master planning, creating a new CDC to address this issue." Other cities see gentrification as something coming to their city soon: "Definitely going to be something that New Bedford is starting to face . . ." However, it is not currently a problem for the city. "I have not heard a loud cry about gentrification . . . It's a double-edged sword right now . . . it's hard to do market-rate housing." Mr. Luongo spoke about how market-rate developers have a hard time getting financing from banks because New Bedford doesn't have a reputation as a place to invest. Instead, the city attracts developers to do affordable housing so they can tap low-income housing tax credits and other public sources. But more affordable housing in the city is not helping to pull New Bedford into a better investor light.

But 62.9 percent of officials felt that there was no evidence of gentrification and were not in the least bit concerned about it. This potentially can be attributed to the overall low median household income and high poverty level for cities included in this sample. While a few cities have begun to see reinvestment in their neighborhoods, most are desperately trying to manage continued decline in population and employment.

How Successful Are They?

Beyond asking the officials about the way they address HI-TOADS, I asked them all how good a job they thought they were doing overall. In assessing Buffalo's efforts, Mr. Cammaratta noted, "we haven't been as impassioned as we should have. On a scale of one to ten, I would give us a four or five." Another planner assigned blame for his northeastern city's lack of success. "We can't be strategic, we don't have a nonprofit development corporation. [The State] DECD [Department of Economic and Community Development] is asleep at the wheel." Attributing her city's failures to other factors, one economic developer admitted, "we're not that good with industrial sites . . . a function of so much, so many up-front costs, rehab, environmental, demotion . . . it's immense."

With respect to the fourteen policies listed in table 4.1, most officials were unable to certify that any one policy caused a successful HI-TOADS redevelopment to happen. Most emphasized the need for a set of policy tools to address HI-TOADS. Several officials were leery of attributing a successful HI-TOAD site redevelopment to a specific policy when there are so many other confounding factors. Many officials expressed their feeling that real estate market conditions can play a larger role in what happens on a site than local government action. Mr. Gower put it best: "We only impact business decisions on a tangent, and that's on a good day." Nevertheless, it appears that in each of the suites of policy tools—zoning, condemnation, economic development, site work, and planning—local government intervention had some impact on HI-TOADS. Examining the way this happens will be a focus of Stage 3 of this research, the case studies.

In order to better understand variation and differences in responses by local officials, I also collected data in the interviews about each official and about their cities. Table 3.2 (in chapter 3) provides descriptive statistics for those variables. How much do the independent variables featured in table 3.2 affect the key dependent variables in this study regarding policies adopted to address HI-TOADS and the extent that local officials see their city as being successful in addressing HI-TOADS?

Three of the independent variables appear to have a significant relationship with whether the city's policies were deemed successful: region, relationship with state government, and relationship with federal government. Due to the small number of cases, only 36 officials in 20 cities that had HI-TOADS, and the fact that only 4 of the 28 officials said that they were unsuccessful, it is difficult to infer much from these relationships. However, a visual inspection of the data reveals an interesting pattern: Officials who used planning and zoning policies were more likely to have stated that their cities were successful in addressing HI-TOADS than those who did not.

The five cities whose officials felt that they have not been successful in addressing HI-TOADS are all in the Northeast, whereas those who felt they are successful are spread evenly among all three regions in this study, the Northeast, Midwest, and South. This was echoed in interviews, where many southern officials claim that HI-TOADS are not a difficult problem.

The extent to which local officials believe that they have been successful is also associated with the level at which they assess their relationship with state and federal government officials. The mean score attributed to the cities' relationship with state officials (on a scale from 1 to 10) was 7.62 for those officials who deemed their efforts to date to be unsuccessful, but 7.81 for the successful ones. The mean score for the unsuccessful cities' relationships with federal officials was 6.90 and 7.91 for successful ones. The importance of strong intergovernmental relations was expected and this finding further supports the need for local officials to maintain good relations with both state and federal officials.

A visual inspection of the cross-tab tables suggests that a potentially important relationship between local action and perceptions of success in addressing HI-TOADS may exist (see tables 4.6, 4.7, and 4.8).

Those officials whose cities adopted zoning policies account for nearly half (46 percent) of those officials who considered their cities to be successful in addressing HI-TOADS, while officials who used zoning accounted for only 25 percent of those who considered themselves to be unsuccessful.[6] Likewise, those officials whose cities adopted planning policies account for more than half (54 percent) of those whose cities were successful, while they account for only 25 percent of those who were unsuccessful. Such differences do not appear in examining cross-tabs for the other groups of policies. In table 4.8, the cities that used both planning and zoning policies (n=5) are examined against their stated success. All five officials whose cities used both planning and zoning considered themselves to be successful (Rochester, Richmond,

Table 4.6

Cross-Tab between Zoning and Success

	Adopted zoning policies	*Did not adopt zoning policies*	*Total*
Unsuccessful			
Count	1	3	4
Percent within Success	25	75	100
Success			
Count	11	13	24
Percent within Success	46	54	100
Total			
Count	12	16	28
Percent within Success	43	57	100

New Haven, and New Bedford). This analysis suggests that the adoption of planning and zoning policies may contribute to whether officials claim their cities have been successful in addressing HI-TOADS. Due to the small number of cases examined, it is not possible to draw any further conclusions. However, these results will inform a new series of hypotheses to be tested in the case study research.

Cross-tab analysis was performed on how cities respond to HI-TOADS by examining the relationship between the eleven independent variables and how many of each of the five classes of policy tools each city adopted. The first class, zoning, was significantly associated with region. Midwestern officials had systematically different policy approaches to development; of the six interviewed, none of them mentioned zoning policies in their efforts to address HI-TOADS, while the answers were more mixed among northeastern and southern officials.

The second class of policies, condemnation, was systematically related to whether intergovernmental relationships matter. Of the 13 officials who said that their relationships with county, state, or federal officials are important to their efforts to address HI-TOADS, all of their cities adopted condemnation policies. In contrast, of those eight officials who did not mention condemnation policies, they were split between five that thought that intergovernmental relations matter and three that did not. This suggests that the use of condemnation is tied strongly to a city's ability to work

Table 4.7

Cross-Tab between Planning and Success

	Adopted planning policies	Did not adopt planning policies	Total
Unsuccessful			
Count	1	3	4
Percent within Success	25	75	100
Success			
Count	13	11	24
Percent within Success	54	46	100
Total			
Count	14	14	28
Percent within Success	50	50	100

well with officials at different levels of government, and that cities that do not have good relationships are less apt to utilize the full range of tools available to them.

Two variables were systematically related to whether a city adopted economic development policies, the third class of policies identified in this study: form of government and whether an official worked as a planner or not. Of the 21 cities included in the study, 15 (71.4 percent) have a strong-mayor form of government, two (9.5 percent) are under a control board, two (9.5 percent) have a council or commission form of government, and two (9.5 percent) have regional governments. The strong-mayor cities are more likely to have adopted at least one economic development policy than the other cities. Ninety-two percent of officials in strong-mayor cities reported having at least one economic development policy, while the results were more mixed for officials with other forms of government (83 percent of officials in non-strong-mayor cities had at least one economic development policy). This finding was expected due to the leadership role that a mayor often plays in bringing in new jobs and businesses and serving as a booster (in an economic development sense) for her city.

Planners normally are engaged in economic development work, so they are expected to discuss economic development policies as a means to address HI-TOADS just as often as other professionals in a city government. However, the cross-tab analysis reveals that planners are less likely to report that their city uses economic development policies than local officials who identify themselves as brownfields coordinators, economic developers, or other positions. For this study, a professional is a planner if he or she is engaged in those activities defined in chapter 2 as planning. All of the interview subjects who spoke about their involvement in HI-TOADS met that criteria. However, it can be revealing to study how people self-identify and whether they see themselves as planners or not. Of the 17 officials who identified themselves as planners, four (23.5 percent) did not mention any economic development policies in the interviews. In contrast, all 21 non-planners mentioned at least one economic development policy. This finding suggests that the interviews may not be the most reliable manner to collect information on policy responses, when large differences in responses emerge based on position. Individual officials may not be capable of reporting fully all of a city's policies. The technique of conducting two interviews in each city allows me to compare answers on policy approaches among planners and non-planners and helps to improve the study's overall reliability. But the extent to which planners are either unfamiliar with or resistant to economic development policies in addressing HI-TOADS deserves further investigation in the case studies.

Region emerges again as a major driver for the fourth class of policies, site work. Where midwestern cities distinguished themselves with respect to their universal failure to adopt condemnation policies, it is midwestern

Table 4.8

Cross-Tab between Planning/Zoning and Success

	Adopted both planning and zoning policies	Did not adopt both planning and zoning policies	Total
Unsuccessful			
Count	0	4	4
Percent within Success	0	100	100
Success			
Count	5	19	24
Percent within Success	21	79	100
Total			
Count	5	23	28
Percent within Success	18	82	100

cities, along with northeastern cities, that take the lead in adopting site-work policies. Ninety percent of northeastern officials and all midwestern officials mentioned at least one-site work policy in their interviews, while only 58 percent of southern officials did so. This may be attributed to a stronger real estate market, actual lower levels of contamination, or perceived lower levels of contamination in southern cities. Without the fear of pollution, less public intervention is needed for Phase I and Phase II environmental site assessment (a major aspect of site work).

The belief that intergovernmental relationships matter to HI-TOADS reuse is associated with cities' adoption of site-work policies. Twelve officials (85.7 percent) asserted that intergovernmental relations matter, and two (14.3 percent) said that they did not. Of those two, one indicated a single site-work policy, while the other indicated four. For the twelve officials who said that relations matter, only three said that their city used one site-work policy, seven said they used two polices, two said they used three policies, and none said they used four policies.

The last set of policies to address HI-TOADS, planning, was systematically related to one variable, the value of grants received in the last year. As the dollar value of grants received by a city for use in HI-TOADS neighborhoods increases, the implementation of planning policies decreases (Spearman's correlation = -0.442).[7] Here, grant monies are seen as a surrogate for the capacity of a city to address its problems. If a city is able to attract external grant monies, then it also tends to have a skilled and well-organized staff with a high level of capacity. By that logic, the cities with lower institutional capacity appear to be more likely to employ planning to address HI-TOADS than other cities. Or seen in the reverse, cities with high institutional capacity do not implement planning policies. Either way, the finding is unexpected and will be explored further in the case studies.

Local Officials' Views on HI-TOADS

"Do local officials recognize HI-TOADS as a problem in the neighborhoods of their cities?" The answer is a resounding yes. While three officials denied having any HI-TOADS in their cities, the remaining 35 officials interviewed could identify at least one HI-TOAD site. Those officials identified sites as HI-TOADS only if they believed that those sites had negative pull on property values at least a quarter-mile away. Most officials thought that

HI-TOADS threaten neighborhood stability and directly cause a wide range of impacts, including crime, arson, dumping, and brownlining.

In answering the second question of my investigation, it is important to note that I cannot generalize my findings to all cities, only to cities that are relatively likely to have neighborhoods with HI-TOADS. For cities not included in the analysis because they ranked low in the multivariate statistical analysis conducted in Stage One, such as Stamford or Las Vegas, I excluded them from the interviews because they are likely have few if any HI-TOADS.

Likewise, while the results presented from the cross-tabulations are interesting and will be valuable in framing the case studies in the subsequent chapters, they are not entirely valid. When conducting cross-tabulations, the validity of the results is highly compromised when any subset (that is, those cities that viewed themselves as successful and adopted zoning to address HI-TOADS) yields fewer than five or ten observations. Many of the cross-tabulations presented in this chapter have five or fewer observations, so any relationship that appears to exist could be spurious.

When discussing HI-TOADS in their cities, the officials recognized that the HI-TOADS were a problem in and of themselves and deserving of public response. Of the officials in the 21 cities that identified at least one HI-TOAD site, all discussed a role for the city government in responding. But the level and extent of public involvement varied across the cities.

Of particular importance was the overwhelming extent to which local officials drew on neighborhood life-cycle theory in formulating their responses to HI-TOADS. With only the single exception of Dayton's asset-based planning, no other officials drew on alternative neighborhood-change theories. The case studies will explore further the dominance of the neighborhood life-cycle theory as a foundation to address HI-TOADS.

"Of those cities that acknowledge HI-TOADS as a problem, what policies do they use to address them?" The officials' responses to this question varied, but I grouped their answers into a set of fifteen distinct policies in five broad categories. Each class of policy tools was employed by 40 percent or more of cities: zoning (40 percent), condemnation (55 percent), economic development (95 percent), site work (90 percent), and planning (55 percent). Because of the widespread adoption of these policies, they are also well known to local officials. However, in the interviews, other policies that cities have adopted to address HI-TOADS emerged: asset-based planning,

interim uses, form-based planning, and historic preservation. Each of these policies offer interesting potential for addressing HI-TOADS and their use will be explored further in the case studies.

These tools were connected to a larger set of city objectives: economic development, community empowerment, and healthy environment policies. Very few officials saw HI-TOADS reuse as a way to achieve community empowerment; they were far more focused on economic development and healthy environments objectives. This orientation away from community empowerment deserves further consideration in the case study portion of the study.

"How successful are those policies?" This second stage of the study began to answer this fourth question. As judged by the responses of local officials, the cities in the sample are quite successful. However, as evidenced by the median number of current HI-TOADS—three in each city—it seems that some work remains to be done. More importantly, the stories the officials recounted of their HI-TOADS suggest to me that there is *much* work to be done. Stories of moribund sites like the Starter Site, a former textile plant on James Street in New Haven, Connecticut, are exemplary. Abandoned years ago, the roughly seven-acre site is seen by local officials as a force that is bringing down property values in the surrounding area: "It's no question, it's a long-term threat." Yet, the City of New Haven has devoted resources elsewhere and the Starter Site remains idle. Its neighborhood remains threatened.

Three other important findings emerged from the interviews. The first was the extent to which some officials appeared committed to adaptive reuse of HI-TOADS while others were simply disinterested in the idea. Another finding was a better understanding of the role that professionals who self-identify as planners play in addressing HI-TOADS. A third important finding was that nearly half of the cities studied did not use planning tools and techniques to address HI-TOADS.

There was a real split between those cities that viewed HI-TOADS as having historic value and those that did not. Officials in Savannah and Richmond spoke about how their main roles with respect to HI-TOADS reuse was historic preservation. One official explained how "protecting the structures" was the best thing they do to address the HI-TOADS problem in their cities. Typical of the other perspective were officials in Scranton,

Pennsylvania, who recounted to me the story of the beautiful, historic Casey Hotel. The multi-story hotel occupied less than half an acre at the corner of Lackawanna and Adams streets in the heart of downtown Scranton, but due to years of abandonment and neglect, it was a textbook HI-TOAD site. With the help of the city, the hotel was demolished and replaced with a parking deck. The project was recounted as a success by an interviewee, yet the loss of the hotel's historic value was not mentioned and not prioritized by city officials.

My methodology of first contacting the senior planner in each city revealed that great variation exists in how cities organize themselves with respect to their ability to plan for HI-TOADS. Most of the cities that have planners working on HI-TOADS were also part of a larger real-estate, community development, or economic development arm of the city. In most cases, these planners are involved heavily with acquiring and redeveloping city properties. In cities that do not fall into this model, the many planners were not interested in talking to me and were quick to refer me to an "economic development" office that dealt with brownfields.

This finding is of great importance, because it reveals that in those cities, people trained and experienced in urban planning are not involved in HI-TOADS. This leads directly to the second finding referenced above, that planning itself is used by only 55 percent of the cities in the study in addressing HI-TOADS. My interviews revealed that those people involved in HI-TOADS were not doing much in the way of long-term, comprehensive planning (in part because they are not planners, in part because that is not their office's mission). As a result, the cities that are addressing HI-TOADS may be doing so through a myopic lens; the kinds of long-range land-use changes they are experiencing (that certainly could result in more HI-TOADS) are not necessarily being thought through.[8]

All of the interview subjects were planners, based on my definition from chapter 2, but only some worked in official planning departments. Those professionals who self-identified as planners often were not involved in HI-TOADS. Self-identification as a planner is correlated closely to one's training in a professional planning degree program. Even though I define all those engaged in planning activities as planners, there is clearly an important difference between those who self-identify as planners and have training in the field, and those who do not.

Planners are trained to see beyond site boundaries. Again and again, I heard from economic development officials that they really don't look beyond the site boundaries in their efforts to deal with HI-TOADS. On the other hand, it may be precisely because economic development officials focus on the site level that they are attentive to HI-TOADS while planners who lack many site-specific tools do not. These findings seem to be critical for this study, because of their implications for the communities and for planning.

Most of the cities included in the study have strong and vital planning departments, in some cases staffed with dozens of professional planners. Yet, when faced with a neighborhood-wide problem like HI-TOADS, roughly half the cities report that they only address site conditions and do not work beyond site boundaries. The half that do plan and act beyond site boundaries tend to hold public meetings, develop neighborhood plans, and then focus public resources into the HI-TOAD site itself (and in some cases complement that investment with neighborhood grants).[9] This represents two competing models for how cities approach HI-TOADS. While the statistical analysis suggests that planning may be correlated with success, this question will be explored further in the case studies.

One caveat is that some officials may have reported to me that they do not do any planning, yet may in fact do it. The semi-structured interview format provided for consistency and reliability, but may have sacrificed validity. Because the role of planning was not a central part of my initial hypotheses, I did not include a direct question about whether the respondent's city, as part of their strategy to address HI-TOADS, was engaged in space-based, forward-thinking activities involving the setting of goals and means to achieve those goals, derived through some level of public involvement. Some officials may do planning yet did not convey that to me. These ambiguities will be central in the case study interviews.

These findings and the answers to the book's key questions now set up a framework for exploring the questions further in three case studies. I studied the impacts generated by HI-TOADS through interviews with residents and community leaders and site visits. While the initial interviews with city officials provided a quick and easy sense of the approaches that cities take to deal with HI-TOADS, the case studies involved a more in-depth examination of the ways that cities act. In addition, I examined the roles of community-based organizations (CBOs) and local residents. In trying to provide a more full answer to the book's fourth question, I obtained empirical evidence to study the changing neighborhood property values where a supposed successful HI-TOADS remediation was initiated. In addition, through interviews with CBOs and local residents, the local officials' assessments of success can be corroborated. Lastly, I explored how the three key findings mentioned above play out in a closer analysis: How does historic preservation enter into discussions about reuse and redevelopment of HI-TOADS? What is the role of the planner and planning, in general, in city government efforts to address HI-TOADS?

[5]

Redevelopment Policy in a
Municipal Coalition City

TRENTON, NEW JERSEY

HI-TOADS offer a unique challenge to local officials. In the first four chapters of this book, I explored the quantity and distribution of HI-TOADS in U.S. cities. I enumerated the major policies and approaches undertaken by local officials in addressing HI-TOADS. In the third stage of the research, I studied local responses to HI-TOADS through a case-study methodology. The statistical analysis and telephone interviews provided a valid and reliable method for understanding how cities address HI-TOADS; however, the results are far from perfect. As discussed in chapter 3, there are a number of limitations to the methodologies employed thus far. To address those deficiencies, I conducted five case studies of cities included in the earlier stages of the research.[1] In this chapter, I will begin by explaining the objectives for this work, how the case study cities were selected, and the general methodological framework for conducting the analysis. The remainder of the chapter will focus only on the Trenton case study. Chapters 6, 7, 8, and 9 will focus on the Pittsburgh, New Bedford, Richmond, and Youngstown case studies, respectively.

One of the key lessons that emerged from the interviews related to the role of both the planning function and planners themselves in the reuse of HI-TOADS. Another important dimension that helped to identify the 21 cities studied was the extent to which they found themselves to be successful in HI-TOADS reuse. Therefore, I used those constructs as a frame from which to select case-study cities for further analysis. While much can be learned from cities that considered themselves to have failed, I decided to focus my attention rather on understanding those cities that have been active (based upon how many policies the interviewees reported they employ to address HI-TOADS), judged themselves to be successful with HI-TOADS reuse, and utilized either planning or zoning policies in addressing HI-TOADS.

By taking this approach, the stories of redevelopment emerge from the data inductively. The redevelopment literature is replete with examples of case-study research that effectively disentangles complex relationship and teases out causality (see Temkin and Rohe 1996; Bright 2000; Keating, Krumholz, and Star 1996; van Vliet 1997). This kind of research is quite different from that presented in chapters 3 and 4, where a set of discrete hypotheses on how cities address HI-TOADS were tested through deductive research. However, even that research had an exploratory bent to it and the questions were open-ended so as to be inductive and inform the direction of the case studies.

Of the fourteen cities that judged themselves to be successful, twelve (86 percent) reported using planning or zoning policies. Among those twelve, I identified those cities that had adopted seven or more total policies to address HI-TOADS and then sought to generate a list of three geographically diverse cities. After generating a preliminary list of case-study cities, I contacted the officials with whom I had spoken during the telephone interviews to assess their willingness to assist me in executing a more detailed case study. Not all officials were willing. I further narrowed my list of case-study cities to places where I felt that I could get cooperation from local officials, which would lend itself to a richer study. This approach produced a nonrandom yet illustrative sample of five cities.

Trenton (with a 2000 population of 85,258), the depressed capital city of New Jersey, has received acclaim from the U.S. Environmental Protection Agency for its brownfields redevelopment efforts. In 2000, Trenton had a median household income of $31,074 (the national average is $41,994) and a poverty rate of 21 percent (the national average is 12 percent) (U.S. Census 2000).

While technically also a part of the mid-Atlantic region, Pittsburgh's (2000 population, 334,563) proximity to the Ohio border makes it in many respects a hybrid northeastern-midwestern city. Pittsburgh's renaissance from moribund steel city to high-tech center over the last two decades makes it a potentially excellent place to study HI-TOADS reuse. Pittsburgh is also a good place to study brownfields because one of the leading scholars of urban environmental history, Joel Tarr of Carnegie Mellon University, has conducted extensive research there.

The devastation caused by plant closing and deindustrialization hurt few regions worse than New England. Therefore, my third case study is in one of New England's great ports, New Bedford, Massachusetts (2000 population, 93,768). Like other small, distressed cities in the region, New Bedford faces homelessness, unemployment, and crime, but has had important successes in recent years with its brownfields redevelopment efforts.

The fourth city is Youngstown, Ohio (2000 population, 82,026), a renowned midwestern city that has turned its misfortunes in recent years to a new ideological front in the war on urban decline. The case study illustrates how this new thinking affects the city's efforts to address its HI-TOADS.

Lastly, I explored the southern capital of Richmond, Virginia (2000 population, 197,790). While annexation has stemmed the city's population loss, racial strife, persistent unemployment, and poverty have ranked Richmond among the most distressed cities in the nation. Despite the indicators, Richmond has had a very different experience with its HI-TOADS over the last decade: By employing a unique set of historic preservation programs and policies, the city has beat the odds and turned around some of its most challenging HI-TOADS.

The case studies were completed through a combination of semi-structured interviews, direct observation, and analysis of housing data. While each method reinforces the others, the bulk of the research relied on findings from the interviews. See Appendix H for a detailed account of the case-study methodology.

Key Research Questions for Trenton

In this third of the three stages of research, I focus on two main issues: What policies do cities use to address HI-TOADS? How successful are those policies? I also pursue five other sub-questions introduced in chapter 1: (1) Did the city's policies fall into one or more of the broad categories identified in the literature review (chapter 2): economic development, community empowerment, or healthy environments? (2) How were HI-TOADS incorporated into the city's planning efforts? (3) Why did the city employ the strategies it employed? (4) How did the city manage the challenge of historic preservation with respect to the reuse of HI-TOADS? and (5) What is the nature of NGO involvement in HI-TOADS?

What was particularly remarkable about my interviews with Trenton officials was that an intriguing contrast surfaced between what appeared to be high city engagement with HI-TOADS, high perceived success, and extremely low indicators of city quality (poverty, low educational attainment, unemployment, and housing vacancy). In my interviews, local officials reported the use of seven unique policies to address HI-TOADS (the highest reported number of policies was eight, the mean was five). Both officials asserted that Trenton had been successful in HI-TOADS reuse. This finding puzzled me because of Trenton's poor performance on a host of indicators of city quality. Is the city as active as it purports to be? Has the city's activism in HI-TOADS reuse translated into the kinds of successes reported to me by the city's planning director and brownfields coordinator? How do people outside of city government view this track record? Do they assess the city as being successful? If, in fact, the city has been successful, what has been the impact in neighborhoods? Why has this success not been reflected in city-wide indicators displayed in census data?

In the remainder of this chapter, I will present a brief history of Trenton and its redevelopment, profile four HI-TOADS in the city, and describe the roles of the city, other public bodies, and nongovernmental organizations in HI-TOADS reuse. Then I will use thematic analysis to answer some of the central research questions of this book.

First settled in the 1600s, Trenton's midway status between New York and Philadelphia made it an ideal center for early trading by European settlers. The city played a strategic role in the Revolutionary War and at one point key leaders considered it for the site of the capital of the fledging country (Cumbler 1989). But Trenton only truly grew to a great manufacturing center with the 1820s advent of canals and railroads, which connected the city with iron mines and coalfields throughout the region (Cumbler 1989). At the peak of the Industrial Revolution, Trenton was well-established as one of the largest cities in the country and one of its industrial giants. But, around the turn of the century, the city's slogan "Trenton Makes, the World Takes" began to lose some meaning, as the city's pottery and iron manufacturing firms began to lose ground to competitors throughout the country and throughout the world (Cumbler 1989).

By the year 2000, the city's total number of manufacturing jobs had shrunk to 3,000 from a peak of 50,000 in the early twentieth century

(Andrews 1999). While some industrial firms adaptively reused their properties for other uses, such as warehousing or assembly, many simply walked away from their plants. This desertion of industrial properties and land is a key ingredient in understanding why HI-TOADS have come to exist in Trenton. Real estate analysts speak of an area's absorption of new uses as a barometer of market strength. Miles, Berens, and Weiss (2000) define absorption as the rate at which new properties can be leased or sold in a given area. It is widely calculated in the real estate industry as net new space occupied less excess vacancies and new construction. Absorption for industrial land in Trenton has been slow in recent decades. While there is a demand for industrial buildings in the greater Trenton region, the existing stock of industrial buildings in Trenton generally does not conform to modern needs (Colliers International 2006; Capasso interview). The multi-story, narrow column spacing typical of an industrial building from the late nineteenth or early twentieth century does not fit the demands of many of today's firms, which seek single-story, wide-column-spacing structures (Burchell and Listokin 1981). This mismatch between the supply of structures and contemporary demands leads such historic buildings to be labeled as functionally obsolete.

While a mismatch of demand and supply for industrial buildings is a challenge for Trenton, other cities have overcome it through proactive adaptive reuse of such buildings for nonindustrial uses. Herein lies Trenton's greatest challenge. While it is understandable that Trenton's industrial building stock is not suitable for industrial uses, why has the building stock not been reused for nonindustrial uses? Part of the answer is common in most postindustrial cities and lies in the presence of unknown and unquantified environmental contamination, the general attitude and risk aversion of industrial property owners, and the lack of strong laws promoting brownfields redevelopment. But in Trenton, the rest of the answer lies in a more fundamental problem the city faces. Trenton did not just experience an industrial decline; it also changed dramatically in racial, ethnic, social, and economic terms over the last half-century. Scholars have shown that when a city loses low-skilled jobs in manufacturing, it then experiences a concomitant slide in other social and economic conditions (Bluestone and Harrison 1982; Keenan, Lowe, and Spencer 1999; Self 2003). Widespread unemployment ensues, which results in increased criminal activity, greater dependency on public assistance, and further disinvestment by industries.

In Trenton, textbook "white flight" also ensued. What had been an integrated high school in 1967 became virtually all African American and Hispanic by 1975. (Race riots sparked by the assassination of Martin Luther King, Jr., in 1968 were a key turning point for Trenton [Cumbler 1989].) While African Americans comprise 51 percent of Trenton's 2000 population, they comprise only 10 percent of the surrounding Mercer County region of twelve towns.[2]

This brief review of Trenton's history provides some context for understanding the challenges the city faces to find new users (both industrial and nonindustrial) for its vast expanse of former industrial lands and buildings. Beyond pure incompatibility and obsolescence, the environmental difficulties of reusing industrial land and buildings for residential, commercial, or institutional uses are particularly vexing. Currently, there appears to be a modest resurgence in residential demand in Trenton. But it is hardly clear that this demand is strong enough to fill up all of the city's HI-TOADS or if it is strong enough to overcome the hurdles of environmental contamination, either real or perceived.

Beginning with the administration of Mayor Doug Palmer in 1990, the City of Trenton has made a concerted effort to direct its programs, policies, and funding toward brownfields redevelopment (Gardner 2001). Under the leadership of Mayor Palmer's Director of Housing and Community Development, Alan Mallach, the city actively began to confront the change in land use described above.

Mr. Mallach was able to push his agenda forward after the State of New Jersey replaced its primary law addressing site remediation, the Environmental Cleanup Responsibility Act (ECRA), with the Industrial Sites Recovery Act (ISRA) in 1993. ECRA was viewed by many as an impediment to site reuse and was blamed for causing only mildly contaminated sites to remain idle (Garbarine 1996). "ISRA sought to promote more privatization of the regulatory process, to reduce the financial burden of site remediation, to streamline the regulatory process, and to reduce DEP oversight" (Gardner 2001, 60).

According to Mr. Mallach (n.d.), "with the enactment of ISRA, the city of Trenton decided to mount an attack on its brownfields" (1). Another key component of ISRA was the creation of the Hazardous Discharge Site Remediation Fund (HDSRF). The fund provided financial support to municipalities and others for conducting environmental investigations and clean-ups.

FIGURE 5.1. Alan Mallach, former City Planning Director, Trenton, New Jersey. Photo
courtesy of Alan Mallach.

"Within two years [of the enactment of ISRA], the city had received eight
separate grants from the HDSRF totaling over $260,000" (Mallach n.d., 1).

The city also took a significant step in its HI-TOADS strategy by creating
the Brownfields Environmental Solutions for Trenton (BEST) Advisory
Council in 1996. For the last ten years, this council of public, private, and
nonprofit officials has met monthly. Serving on the council are federal,
state, county, and local government officials, as well as representatives of
academia, real estate, engineering, and planning. One BEST member explained that "the city updates us on every property under redevelopment."
He went on to say that the "city uses it to get input from a variety of different people on setting priorities."

During the two decades from 1980 to 2000, Trenton lost 6,721 residents
(7.3 percent) (U.S. Census 2006). The city went from 44 percent African-
American and 8 percent Hispanic in 1980 to 53 percent African-American
and 22 percent Hispanic in 2000 (U.S. Census 2000). During that period, the
city's white population dropped from 49 to 25 percent (U.S. Census 2000).

Mean housing value in the city grew from 1980 to 2000 from $23,197 to
$29,362.[3] While a sizable rise, it reflects boom levels of real estate growth

occurring throughout the nation more than an increase in the residential desirability of Trenton. In the state of New Jersey, median housing values rose from $119,200 to $170,800 during the same time period. Likewise, median gross rent values tripled in Trenton during the twenty-year period, from $208 per month to $616 per month. Average annual household income also grew during this time, more than doubling from $14,351 to $38,511.[4] But statewide, annual household income grew by even more, from $19,800 in 1979 to $55,146 in 1999. Another indicator of wealth, homeownership, slipped in Trenton from 58 percent in 1980 to 47 percent in 2000.

This demographic data paints a picture of a distressed city, in both economic and social dimensions. Trenton scored high on the very indicators associated with the presence of HI-TOADS: unemployment, housing vacancy, and poverty levels, among others. The central question for this case study is, which of the following three scenarios happened?

a. To what extent have HI-TOADS continued to drag down the city?
b. Through their reuse, have HI-TOADS helped revitalize neighborhoods?
c. Neither has happened.

For this and the other case studies, the focus is on all existing and all former HI-TOADS that were redeveloped from 1995 through 2005. By maintaining a narrow time frame for the analysis, I am able to closely study local policy responses to HI-TOADS in that ten-year timeframe and their results.

Based upon the telephone interviews I conducted, it appeared that there were six HI-TOADS in Trenton during the study period. This case study confirmed that there were eight HI-TOADS in Trenton, but a slightly different set than the ones I learned of earlier. In the interviews, I learned about the Roebling Complex, Crane Pottery Site, Assunpink Greenway, American Bridge and Steel Site, Prospect/Remmington/Oakland Streets Area and the Magic Marker Site. During the case studies, I determined that the American Bridge and Steel Site was redeveloped more than ten years ago and that the Prospect/Remmington/Oakland Streets Area did not qualify as a HI-TOAD site (due to local officials' assessments that it did not represent a drag on neighborhood property values). In the case-study research, I also learned about four additional HI-TOADS: the Thropp Site, PSE&G Site, Trenton Waterworks, and the Champalle Site.

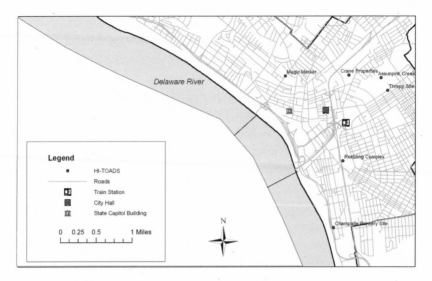

FIGURE 5.2. HI-TOADS in Trenton, New Jersey.

Four HI-TOADS—Magic Marker, Crane Pottery, Roebling Complex, and Assunpink Greenway—help tell the story of HI-TOADS redevelopment in Trenton, warts and all (see figure 5.2 and table 5.1). Within the framework of the three urban policy approaches, economic development, community empowerment, and healthy environments, I identify the overriding policy orientation for each site.

The Trenton HI-TOADS

Magic Marker Site

In the mid-1930s, a battery manufacturer commenced operations at the Calhoun Street site in northwest Trenton (U.S. Environmental Protection Agency 1998). Decades of toxic operations ended in the 1980s, when the site was transferred to Magic Marker Industries, world-renowned manufacturer of felt-tip pens (U.S. Environmental Protection Agency 1998). The site was shuttered in the 1980s and a legacy of environmental contamination soon was exacerbated by the presence of abandoned buildings in a largely residential neighborhood. A paper by Mayer and Shaw (1997) explained the consequences of the closure: "a general sense of decline has affected the overall neighborhood and along with the loss of jobs in the area, there has been a drop in city services and in increase in the drug trade" (7). A mix of trespassers, hazardous materials, and arson made the Magic Market Site a prime target for redevelopment. While the city was ramping up its new brownfields program in 1993, the New Jersey Department of Environmental Protection (NJDEP) was first to become involved directly in the Site. Through its research office, and under the direction of Judy Shaw, the NJDEP asked, "Could a grassroots approach to remediation, involving neighborhood residents, more effectively accomplish what the state alone could not solve?" (Mayer and Shaw 1997, 10). NJDEP then issued a solicitation for a nongovernmental organization to attempt such a grassroots approach. Isles, Inc., of Trenton was selected. Working closely with NJDEP, Isles began a concerted effort to galvanize local support for demolition, remediation, and redevelopment of the Magic Marker Site. One result was the formation of the Northwest Community Improvement Association, a neighborhood group focused on the community health threats of the site and planning for the reuse of the site and general neighborhood improvements. According to Trenton's Brownfields Coordinator, J. R. Capasso, "they were involved in clean-up issues and exploring reuse possibilities." Trenton's Principal Planner, Bill Valoochi, explained the role of local residents: "We didn't have any idea of what we were going to do . . . seven or eight years ago . . . we went to people's living rooms and sat down with them." Another result of the grassroots effort was that further funding was generated from state and federal sources to address not just Magic Marker, but other sites throughout Trenton. In 1999, the city demolished the derelict structures at Magic Marker. According to a former official, the city used its entire building demolition budget just on Magic Marker that year. Plans have been adopted to construct two dozen units of affordable housing and a neighborhood park at the site. Environmental remediation work is ongoing, but city officials do not have all the money needed. They are continuing to seek assistance to fund the remediation from outside sources, including the state, county, and federal governments.

In 2007, an unimproved site is surrounded by imposing fencing punctuated with warning signage posted by state and federal regulators (see figures 5.3 and 5.4). Due to the busy activity of remediation and a general sense that the site is secure, it appears that it is no longer affecting property values more than a quarter-mile away and therefore is no longer a HI-TOAD site.

Table 5.1

Summary Information for HI-TOADS in Trenton

Site name	Size	Location	Historical use	Proposed use	Status of redevelopment efforts (as of 2006)
Thropp site	2 acres	960 E. State St.	Foundry	Park and residential uses	City is seeking funding to conduct remediation and redevelopment.
Crane Pottery site	11 acres	North Clinton Ave. and Dewey Ave.	Pottery manufacturing	Light industrial	Redevelopment resulted in four new industrial buildings.
Magic Marker site	7.5 acres	467 Calhoun St.	Battery and felt tip pen manufacturing	Mixed use residential, commercial, and park uses	Environmental remediation of the site is underway. Structure demolished in 1999.
Roebling complex	45 acres	Along Rt. 129 between Hamilton Ave. and Dye St.	Steel cable (wire rope) manufacturing	Mixed use residential, commercial, office, entertainment, education, and other uses	Roughly half the site has been reused. Plans are underway to redevelop the remaining portion of the site.
Champalle site	2.4 acres	Intersection of Lafor St. and Lamberton St.	Brewery	Residential uses	Structures demolished in 1998. Plans are underway to construct new housing.

Assunpink Greenway	30± acres	Beginning at intersection of E. State St. and Monmouth St. and extending along the Assunpink Creek to Mulberry St. and St. Joes Ave.	Wide range of industrial, railyard, and junkyard uses	Park and recreational uses	City has begun acquiring parcels and has developed a master plan for the project.
PSE&G site	9± acres	New York Ave.	Electrical generation	Entertainment	PSE&G has completed clean-up of the site and is preparing for new use.
Trenton water works	1± acre	Humboldt Sweets	Bottling works	Multi-family housing and commercial	City has acquired title to the property and is working toward a redevelopment.

SOURCE: City of Trenton (1999); personal interviews with local officials.

FIGURE 5.3. Magic Marker Site: In the process of remediation—looking west from Calhoun Street.

The rallying effort of city, state, federal, and community organizations appears to have diminished the deleterious impact Magic Marker once had
on its neighborhood. In terms of the policy triad of economic development, community empowerment, and healthy environments, the city's efforts at Magic Marker largely fell into the community empowerment
model. The city has partnered with a variety of organizations in support of
community-defined objectives (in this case the demolition, remediation,
and reuse planning for the site).

While talking to local officials and community leaders shed much
light on HI-TOAD redevelopment efforts in Trenton, I now turn to
quantitative data to scutinize these preliminary findings further. The
U.S. Census collects vast amounts of data on housing values at a remarkable level of detail and scale. While this information could be used to
prove that HI-TOAD reuse changes neighborhoods, I did not attempt to
do such a thing. Here, I present the results of an analysis of neighborhood changes in Trenton simply to frame the qualitative findings of the
case chapter (see Appendix H for a detailed discussion of the census
analysis conducted).

The census data ends up telling a similar story to that presented
above. With the building demolition in 1999, Magic Marker ceased to be

a HI-TOAD site. The combined efforts of public and private bodies had generated sufficient attention by that time that the site appears no longer to generate any negative externalities; if anything, it may have generated positive externalities due to the hopefulness in the community that a new development was imminent. Therefore, the key time period for this analysis is from 1990 to 2000. During that time, mean housing prices in the Magic Marker tract grew 35 percent, a major contrast with the 17 per-cent fall in prices citywide.[5] Median gross rent grew 43 percent, roughly the same as the 46 percent growth in gross rent citywide. The active in-volvement and financial commitment by multiple public sector actors in the remediation and reuse of the site primarily took place between 1990 and 2000. While population in the Magic Marker Census Tract fell by 16 percent during that period, housing prices and rental rates rose. Mr. Ca-passo remarked "it's taken so long . . . it's almost like an infection where you get pockets of good stuff done . . . there is now some momentum in the neighborhood."

In order to check these results, I also consulted residential real estate sales data collected by the company Trend MLS. That data was provided for the zip code hosting Magic Marker and for the years 1998 and 2005. During

FIGURE 5.4. Magic Marker Site during remediation, looking north from Calhoun Street. Deep swales have filled with rain water.

that time period, an explosive growth occurred in the wider regional housing market that was driving up interest in Trenton. Citywide, the number of transactions soared from 175 in 1998 to 1,123 in 2005, with mean sales price growing over 100 percent. In the Magic Marker zip code, the number of transactions increased from 95 to 370 with mean sales price climbing 80 percent. More than anything else, these results show that wider trends toward the mid-2000s seemed to be ameliorating any lingering downward price effects the Magic Marker site had on its neighborhood. By 2005, the consensus was that Magic Marker no longer met the HI-TOAD site definition and that the city and state efforts were successful.

Crane Pottery Site

Trenton's growth from a small town to an industrial giant of the late nineteenth century was due primarily to its pottery industry (Cumbler 1989). At one time, Trenton-made ceramics competed well in the world market. Today, little remains of that clayworks greatness. Estimates are that fewer than 100 jobs in the field of pottery exist in Trenton (Capasso interview 2006). One such pottery firm, the Crane Company, operated at an 11-acre North Clinton Avenue site for over forty years. By 1970, operations ceased at the site and in 1972 the structures were demolished (U.S. Environmental Protection Agency 2006b).

The site was one of the four initial sites studied by NJDEP and Isles, along with Magic Marker, the Thropp Site, and the Champalle Brewery Site. That early attention, combined with a pre-existing interest by local officials, served as a catalyst for the partnership between the EPA, NJDEP, and John Wiley & Sons of Trenton that resulted in the remediation of the site and the construction of a candle factory in 1998. Shortly thereafter, six additional businesses located at the site for a combined 72 full-time and 11 part-time jobs (NJDEP 2006b; see figure 5.5). To recognize this success, the EPA awarded the efforts of the city, Isles, NJDEP, and the developers, L & F Urban Renewal Properties with one of its prestigious Phoenix Awards for excellence in positive community impact (NJDEP 2006b).

One unique attribute of the Crane Site project, according to local officials, was the way that the large, hulking property was subdivided to meet the current needs of small business better. One BEST member explained that "you will be more successful if you attract smaller businesses. Would a private developer do

FIGURE 5.5. Crane Site: New warehouses along North Clinton Avenue.

that? I don't know if they would take that kind of risk. The city has stepped in and taken a risk, they were willing to go down that path."

The Crane initiative was classic economic development. Through a City Hall–driven process, the site was acquired, demolished, remediated, and redeveloped for the express purposes of economic development. The fact that some surrounding neighborhoods may realize a benefit because a HI-TOAD site no longer exists is purely incidental. The drivers for city policies at the site were determined by the economic development model.

The key date of interest for the Crane site is 1998, when the first new buildings were opened. The effort to investigate and remediate the site had gone through various fits and starts but really got underway in 1990 and 1991. That is, prior to 1990, the site was viewed as a HI-TOAD site. And after 1998, it undoubtedly was not a HI-TOAD site. As greater commitment to a remediation and redevelopment began to appear to local residents and businesses, the negative externalities of the site surely diminished from 1990 to 1998. But because the data I am using is the decennial census, I am particulary interested in what changed in the Crane neighborhood from 1990 to 2000, relative to Trenton as a whole.

As indicated in table 5.2, property values and gross rent in the Crane tract changed at approximately the same rate from 1990 to 2000 as did the

Table 5.2

Property Value and Demographic Data for HI-TOADS and the City of Trenton, 1980 to 2000

	1980	1990	2000	Percent change		
				1980–1990	1990–2000	1980–2000
City of Trenton						
Mean housing value ($)	9,407.87	30,404.64	25,163.23	223	-17	167
Median gross rent ($)	208.00	422.88	616.54	103	46	196
Percent rental units	42	44	49	5	10	15
Average household income ($)	14,351.25	29,376.11	38,510.97	105	31	168
Population	92,124.00	88,675.00	85,403.00	-4	-4	-7
Percent African American	44	47	53	6	13	20
Percent Hispanic	8	13	22	70	66	183
Crane Tract						
Mean housing value ($)	2,635.76	10,271.04	7,303.85	290	-29	177
Median gross rent ($)	175.00	350.00	471.00	100	35	169
Percent rental units	61	66	35	8	-48	-44
Average household income ($)	10,897.59	31,905.29	24,306.80	193	-24	123
Population	2,352.00	1,777.00	1,435.00	-24	-19	-39
Percent African American	96	97	83	1	-15	-14
Percent Hispanic	5	2	12	-54	486	168

Magic Marker Tract

Mean housing value ($)	4,039.35	12,098.97	16,287.49	200	35	303
Median gross rent ($)	219.00	410.00	587.00	87	43	168
Percent rental units	53	52	47	–1	–9	–10
Average household income ($)	12,354.82	25,243.70	34,254.91	104	36	177
Population	4,221.00	3,494.00	2,944.00	–17	–16	–30
Percent African American	94	95	95	0	1	1
Percent Hispanic	3	5	4	51	–22	17

Roebling Tract

Mean housing value ($)	7,000.00	20,467.32	16,250.00	192	–21	132
Median gross rent ($)	241.00	496.00	614.00	106	24	155
Percent rental units	36	46	55	27	19	50
Average household income ($)	14,615.26	30,610.06	36,986.42	109	21	153
Population	2,332.00	2,272.00	2,367.00	–3	4	2
Percent African American	7	13	28	99	115	328
Percent Hispanic	24	39	47	66	20	99

entire city. Homeownership rose, absolutely and relative to the city, from 34 percent in 1990 to 65 percent in 2000, but incomes fell during the same period. The percentage of Hispanics in the tract increased 486 percent between 1990 and 2000 (from 37 to 175 persons), compared with 66 percent increase in the entire city.

As at Magic Marker, I collected residential real estate sales data for the Crane zip code area for 1998 and 2005. And as I found at Magic Marker, the rapid growth in real estate prices regionwide pushed the total number of sales and the mean sales price to record levels in the Crane zip code area. For both number of transactions and mean sales price, indicators show a growth at roughly the same rate in the Crane area as in the city as a whole (530 percent growth in transactions and 91 percent growth in mean sales price during the 1998 to 2005 time period).

This statistical analysis clearly paints a mixed picture for understanding what the impact of the Crane reuse may have been on the neighborhood. More than anything else, it tells the story of a steep rise in the percentage of Hispanics in the neighborhood and little change in the growth of property values and rental rates. The demographic change appears to have little to do with the Crane reuse. The real estate sales data showed robust recovery of neighborhood real estate values; however, the reasons for property value and rental rate stagnation from 1990 to 2000 is unclear. In studying HI-TOADS, the greatest concern is not with gains in property values but rather with eliminating a factor that could be contributing to a drag on values. To understand fully the extent to which the "drag" effect was eliminated at the Crane site, I need to rely more heavily on my other research tools.

During my direct observation of the Crane site, I witnessed two contrasting images: one was several rats and the other was a new self-storage business being built across the street from the site. The rats are a marker for poor neighborhood conditions, likely attracted by decaying garbage that I observed on the property. However, the self-storage business is a marker of a rising neighborhood, where private investors are willing to put money into a place and reap rewards. In describing the residential area across the street from the Crane site, one local official commented, "this neighborhood was bad and still is bad." Perhaps he was correct. While the Crane site will contribute positively to the city's property tax revenues and has produced new jobs, ultimately the neighborhood needs more than the conversion of a HI-TOAD site in order to see any substantive improvements.

Roebling Complex

At one point the epicenter of industrial Trenton, the sprawling 45-acre Roebling Complex was shuttered during the 1970s and 1980s. For most of the 19th and 20th centuries, until the complex began to close, Roebling workers made most of the steel cable used in America's suspension bridges (City of Trenton 1999). As buildings were emptied, plans began to take shape within the community to reoccupy the site. A visitor to the site today would see an arena, offices, a night club, a school, senior housing, and retail stores (see figure 5.6). Today, more than half of the 45 acres have been reused or redeveloped. The new uses are a central component of what one city official called Trenton's search to "find a niche in modern economic times" (Alan Mallach, quoted in Garbarine 1997).

In 1985, the now-disbanded Trenton Roebling Community Development Corporation (TRCDC) took the lead in developing a plan to consider the future of what had quickly become a HI-TOAD site at the Roebling Complex. That original plan later was seized upon by a series of city, county, and private developers as the piecemeal reuse and redevelopment of the site proceeded for more than two decades. As one former city official put it, they "weren't much of a force overall, but they were important originally." He further explained how critical the TRCDC was in the adaptive reuse of many of the historically significant structures at the complex: "they got the idea of historic preservation being part of the project into people's minds."

Around 1995, the TRCDC partnered with a developer to build the Roebling Market and several other projects on Block 1. Several of the Block 1 projects adaptively reused existing structures, including the Roebling Market (see figure 5.6). The project used a combination of state, federal, and city funds and was anchored partially by the development of the New Jersey Housing Mortgage Finance Agency's offices.

According to Mr. Capasso, "the county has taken the lead on a lot of these properties." With a heavy dose of public funding, officials constructed an arena for Trenton's minor league hockey team on Block 5 in 1999 (Garbarine 1997, 1999). The city used its condemnation powers to take Block 3 and are using it for parking to support the arena (see figure 5.7). When I asked local officials about the remaining undeveloped, idle pieces of the Roebling Complex, for which the county is allegedly spearheading further developments, they responded that "they've been floating plans for years."

The most-recent plan was for a fledgling special effects company, Manex, to construct a studio on Block 3. While the proposal generated much fanfare, Manex ultimately did not have the necessary financing and the deal died (Hanover 2006).

While the Trenton Roebling CDC played a key role early in the Roebling reuse, the policy orientation of both city and county efforts was defined largely by economic development, as at the Crane site. The City of Trenton's efforts early in support of the TRCDC's partnership in the development of the Roebling Market indicated a community empowerment orientation, but in recent years, the city's aggressive efforts to recruit outside firms has been indicative of economic development.

From 1990 to 2000, during the period of most active reuse and redevelopment at the Roebling Complex, the mean housing value of the surrounding census tract fell 21 percent while median gross rent climbed 24 percent. For both indicators, the Roebling tract performed worse than the city as a whole. That poorer performance may be confounded by other variables beyond the redevelopment of the Roebling Complex. During that period, the percentage of owner-occupied housing units fell (absolutely and relatively to the city as a whole) and the racial mix shifted dramatically from 52 percent

FIGURE 5.6. Roebling Complex: Adaptively reused Roebling Market, an active shopping center.

FIGURE 5.7. Roebling Complex: Historic warehouse on Block Three and the Manex water tower.

African American and Hispanic to 75 percent African American and Hispanic. The broader changes occurring in the neighborhood, the causes of which are not fully understood, may have skewed the real neighborhood-wide benefits of the Roebling development. City and county officials are unanimous in their approval of the work done to date and its positive external impact.

Assunpink Greenway

In 1999, Hurricane Floyd devastated vast stretches of Trenton when the Assunpink Creek flooded. One former official explained that "[almost] everything was underwater" in the creek's floodway. The Federal Emergency Management Agency (FEMA) provided funding to the City of Trenton to develop a hazard mitigation plan for the floodway. Beginning four years earlier, city officials had begun thinking about a redevelopment strategy for the area around the creek, but after the hurricane, the effort "took on steam."

Land use along the floodway has been primarily industrial for more than one hundred years. New Jersey Transit operated a now largely defunct railyard along the floodway. A New Jersey Department of Corrections facility undergoing closure occupies almost 10 acres along the creek. The rest of the floodway features several abandoned industrial structures, a junkyard, and an existing city park, George Page Park. In my interviews with public officials, there was consensus that the entire floodway meets the definition of a HI-TOAD site. I was able to confirm this assessment during my visit to the site. The quality of the housing stock diminished greatly as I approached the floodway. Trash was strewn about and vegetation was overgrown throughout (see figures 5.8 and 5.9). The quantity of abandoned buildings directly along the creek was astounding. The floodway areas were unsupervised and ideal locations for criminal activity. In fact, the morning that I visited the floodway, the city's major newspaper ran a story of a murder that occurred in the floodway (Shea 2006).

As a result of the FEMA-funded hazard-mitigation planning, the city designated a 30-acre area roughly 200 feet north and south of the Assunpink Creek all the way from East State Street to Saint Joes Avenue as the Assunpink Creek Greenway. Shortly afterward, the EPA had assigned one of its own employees, Leah Yasenchak, to work in Trenton's City Hall to

FIGURE 5.8. Assunpink Greenway: View along Northeast Corridor rail line looking west. Overhead rail power lines are on the right.

FIGURE 5.9. Assunpink Greenway: Typical abandoned buildings and broken fences along the greenway.

support brownfields redevelopment.[6] Upon arriving, the city asked Ms. Yasenchak to take the lead on the greenway project. Thus, beginning in 1999, Ms. Yasenchak championed the refinement of a master plan for the greenway and coordinated the city's slow and steady acquisition of parcels, both through negotiation and through condemnation under the New Jersey Redevelopment Act.

This top-down process has involved virtually no citizen outreach or participation. According to one former official, "community groups did not play a significant role" in the Assunpink Greenway planning. But, despite this weakness, the area-wide approach to the multiple abandoned and semi-abandoned sites is quite innovative. Success is hard to judge at the Assunpink Greenway. With the exception of a few scattered city-initiated demolitions, to date little in the way of physical improvements have been accomplished in the area. However, the development of a master plan and steady progress on acquiring sites has begun to have a snow-ball effect, according to those I interviewed. For many in Trenton, the greenway concept is the only reasonable land-use strategy for the floodway. The fear of future flooding makes a park and recreational use the wisest use, but because such improvements will not directly generate greater tax revenues or jobs, obtaining external funding has not been easy for the city.

In contrast to the economic development and community empowerment policies adopted by the city with respect to the other HI-TOADS in Trenton, the Assunpink Greenway efforts are illustrative of healthy environments policies. The intentional, planned deintensification of land uses and systematic depopulation of an area is textbook healthy environments. Interestingly, this policy orientation was driven largely by a natural disaster, Hurricane Floyd in 1999. Also, interestingly, the greenway is stalled in the planning stages, with much work to be done to implement it fully.

I did not examine census data for the Assunpink Greenway site, so it is not included in table 5.2. I omitted it from that analysis because it traverses five census tracts, while the other five HI-TOADS are each contained within a single census tract. Both the Thropp and Crane sites are included within those five tracts. Therefore, it would be difficult to isolate the effect of the greenway on housing values and rental rates apart from the effect of the other HI-TOADS. In addition, because the project is currently in an embryonic stage, it is hardly fair to test the success of the city's intervention at this point.

The real estate sales data and census data for the HI-TOADS examined in Trenton appear to support the conclusions drawn from the interviews and direct observation of conditions at each HI-TOAD site. Census tracts that are host to the Crane site and the Roebling site performed worse than the city as a whole for the key property value variables. However, the Magic Marker's conversion from a HI-TOAD site to a non-HI-TOAD site may have been a contributing factor to the superior performance in that tract. Overall, the analysis reveals that each of the census tracts examined are some of the poorest, lowest property value neighborhoods in the city and therefore the direct impact of the HI-TOADS reuse may not be detectable at the coarse level of a census tract or the infrequent measurement of a decennial census. In the following sections, the question of success will be pursued further, drawing on measures other than property values and rents.

How Are HI-TOADS Addressed in Trenton?

"If we don't do it, nobody will" is the mantra of the City of Trenton's public actors in brownfields redevelopment. Mr. Capasso explained, "the buck stops with us . . . it's really impossible for the private sector to do it alone." "The position that I took is that the City of Trenton has to address these sites, that if the city didn't, no one would" remarked a former official.

One former city official explained the economics of it: "Developers will take on responsibility for environmental contamination in direct proportion to the extent of profit they calculate . . . [In the early 1990s,] there was hardly a site in the city where developers would take on that responsibility. We would either do or finance the remediation to make sure the projects would happen."

One of the hallmarks of the city's approach was to first gain control of the HI-TOADS. "If you can't control the property, you can't control what happens in the neighborhood," commented a former city official. The "city was aggressive in taking ownership of the sites," remarked another former official. At several HI-TOADS, the city acquired an interest in the property either through amicable negotiation or through eminent domain. Of the eight HI-TOADS examined in this case study, the city acquired some level of ownership in four. That activism on the part of the city appears to have been a key ingredient in getting those four projects moving toward redevelopment.

Another important city policy in the reuse of HI-TOADS was the creation in the late 1990s of a position of brownfields coordinator housed within the city's Department of Housing and Economic Development. The position was installed around the same time that Trenton was designated a Brownfields Showcase Community by the EPA.

A slew of new funding commitments from EPA and other federal agencies followed, as well as the assignment of a full-time EPA employee based in Trenton. The new brownfields coordinator helped to manage many of these new sources of funding and was charged with securing more external funding. As I discussed in chapter 4, Hartford, Connecticut, officials believed that the elimination of the position of brownfields coordinator was a contributing factor in their city's failure to address HI-TOADS successfully. Likewise, Trenton's commitment to a full-time employee working exclusively on brownfields has helped to sustain much progress on the city's HI-TOADS. Mr. Capasso and Michelle Christina (his predecessor) worked directly on nearly all of the city's HI-TOADS, serving as central clearinghouses of information for public officials and private developers.

In my national telephone survey of local officials, I learned that most officials viewed a "shovel-ready" policy approach as key to their redevelopment of HI-TOADS. But, there is a tension between "shovel-ready" and "adaptive reuse." In Trenton, the city aggressively pursues "shovel-ready," but historic preservation has been undertaken. At the Roebling Complex, the city went along with a local CDC's plans for adaptive reuse of some structures, but maintained an active demolition program for others. At both Crane and Magic Marker, the abandoned structures were not perceived to have any historic value and community support for demolition was high. "A lot of the older buildings in Trenton have a problem, basically [that] masonry building type, brick-works structures, if they sit open, exposed to the elements, they become destabilized."

Rather than seeing themselves as "a preservation force to be reckoned with," as one southern official proclaimed during my telephone interviews, Trenton officials were ready to demolish structures that they saw as obsolete for modern industrial purposes. But unlike the demolish-and-wait approach advocated by some officials, Trenton first devised neighborhood plans for most HI-TOADS and then, after deciding on a reuse, went about preparing for that goal. Mr. Mallach (n.d.) wrote about Trenton's planning strategy, "too often, municipalities, as well as private owners allow the environmental issues to drive their thinking about the property, rather than subordinating those issues to the major one: namely, what can (and should) the property be used for?"

Starting from that envisioned land use, Trenton devised plans and protocols for various situations to demolish structures, investigate environmental contamination, conduct environmental remediation, market the property, and subsidize a redevelopment. Taking the property from soup to nuts, financing from both city and external monies, the city successfully redeveloped hundreds of acres of brownfields. Of the eight HI-TOADS examined, the Crane Site and Magic Marker Site are successes, while the Roebling Complex is a mixed success. Remaining to be addressed and for neighborhood impacts to be minimized are the Thropp, Champalle, Assunpink Greenway, PSE&G Site, and Trenton Waterworks.

While the focus of this book is on the local public policies of HI-TOADS redevelopment, county, state, and federal action is clearly an important part of Trenton's story. On each level of government, a concerted effort has been made to address HI-TOADS.

According to my interviews with local officials, Mercer County has historically played a background role in most urban redevelopment in Trenton, providing modest financial support and boosterism for scores of projects. That role shifted when the county took on a new role in the late 1980s as a real estate developer. The Mercer County Improvement Authority was the lead agency for a new baseball stadium, office, and entertainment complex along the Delaware River. The county then moved on to take a lead role in the development of the Roebling Complex because of that site's proximity to its Riverfront. While the county deserves some credit, the Roebling Complex is still not fully reused and my statistical analysis of property values shows that the benefits to the neighborhood from the project are mixed.

Being the capital of the state, Trenton has been somewhat of a favorite among state agencies. Two agencies play the most important roles in HI-TOADS reuse: the New Jersey Department of Environmental Protection (NJDEP) and the New Jersey Department of Community Affairs (NJDCA), where the Office of Smart Growth is housed.

While state officials generally are pleased with how HI-TOADS are addressed by the city, they recognize a limit to how far local resources can go. Where NJDCA regularly disperses funding to the city for a myriad of economic development programs, those funds are rarely available for the kinds of planning, site preparation, remediation, or development funding that Trenton seeks for its HI-TOADS. For those funds, Trenton has looked to NJDEP.

The challenge of engaging NJDEP is that the agency will provide support for reuse of a HI-TOAD site only when either a city or a private entity is ready to clean a site. "If nobody steps to the plate, then nothing happens . . . the site would have to be grossly contaminated to be at the top of the department's list" explained Kristin Pointin-Hahn of the NJDEP. She continued, they just "kind of sit back, it's unfortunate . . . it's a resource issue, a budget issue." But the NJDEP did develop the Brownfields Development Area (BDA) program in response to this exact challenge.

The BDA program targets a neighborhood with an agglomeration of brownfields sites. The point of the BDA is "to do multiple brownfields at once because of their very HI-TOADS nature" explained one state official. Of the 15 BDAs designated throughout the state, two are in Trenton, one around Magic Marker and the other along the Assunpink Greenway.

The state uses the BDAs to prioritize funding for neighborhoods in the greatest need. As Ms. Pointin-Hahn explained, "we're supposed to treat it as a priority case." In addition, the NJDEP has arranged for "a single case officer for the whole city," according to a member of Trenton's BEST Council.

But it was NJDEP's early study of the role of community organizations in promoting the reuse of contaminated properties that brought attention to the Magic Marker site. Arguably, that research project served as the catalyst for the successful application to EPA for a Brownfields Pilot Grant, which in turn led to a series of other grants and loans to support HI-TOADS reuse. According to one state official, "we picked the sites that were the biggest blight in the community, these qualified as HI-TOADS before anyone knew what a HI-TOAD was."

Here, the state served as a source of innovation, a true rarity in today's federalist system. One state official attributed this unique contribution to the fact that NJDEP employees feel that their agency is "known nationally as one of the forerunners of environmental policy and regulation . . . NJDEP has prided itself on breaking new ground." The impact this had on HI-TOADS in Trenton is no accident, given that the city is the state's capital and NJDEP's headquarters are located there.

While county and state governments have played an important role in support of local efforts at HI-TOADS reuse, it is the strong federal commitment to Trenton that has lent the city great credibility. Over the last decade, Trenton has consistently beat out its peer cities in receiving federal aid. The federal government has bestowed one award after another to Trenton, from the initial Pilot Grant, to Showcase Community, to a series of EPA Phoenix Awards. The City of Trenton has been profiled on scores of EPA publications and websites as a serious place to do brownfields redevelopment. This partnership between EPA and the City of Trenton extended on a personal level when EPA assigned one of its own employees, Leah Yasenchak to work full-time in support of Trenton's brownfields program and to be stationed in City Hall. The fact that Ms. Yasenchak was first tasked with working on a HI-TOAD site, the Assunpink Greenway, speaks volumes.

Other federal agencies also supported HI-TOADS redevelopment through a variety of funding programs at various sites. Funding came from agencies including the U.S. Department of Housing and Urban Development (HUD) and the Economic Development Administration. But the EPA has taken the lead in providing technical and financial support for HI-TOADS reuse project in Trenton.

In studying the eight HI-TOADS in Trenton, five community groups have emerged as part of the tale of HI-TOADS reuse: Isles, Inc., Crane Site Committee, the Northwest Community Improvement Association, South Trenton Area Residents (STARS), and the Trenton Roebling Community Development Corporation (TRCDC). In addition, the city's BEST Council provides an opportunity for other members of the nonprofit sector to discuss and contribute to the development of city policies related to HI-TOADS The private sector, in close partnership with the city, has played a role in HI-TOADS reuse efforts.

In explaining why so few community groups are active in HI-TOADS projects, a state official said, "the community organizations that have formed to lead, too often, have not had the capacity to handle this kind of work." Due to the complexity of environmental remediation and the intricacies of the political process related to development in the city, only one well-funded, highly trained community group has been truly successful in HI-TOADS reuse work in Trenton: Isles, Inc. While the Crane Site Committee, the Northwest Community Improvement Association, and the TRCDC were part of the process, they did not play an instrumental role in forging ahead the reuse. Rather, each played a more minor, yet important role in shaping the final land use of each site.

In Gardner's (2001) study of brownfields redevelopment in Trenton, she found that "although the city has some strong community organizations, they are involved only to the extent that civil administrators allow" (136). In describing the city's brownfields program she goes on to report: "Although the program appears to be participatory, the local political culture has fostered a governing style that is more paternalistic than cooperative. Benevolent city officials and city planners believe they know what is best for the communities and express frustration when residents disagree with their plans" (137).

Private, for-profit developers also have played an important role in the city's effort to address HI-TOADS. They have partnered with the city in all of the projects done to date, including the Crane site and plans for the

Magic Marker site. Only in 2006, is a private developer, the K. Hovnanian Company, looking to redevelop a HI-TOAD site independent of massive public subsidy. Hovnanian is under contract to acquire the Champalle Brewery site to redevelop it into market-rate housing (Parker 2004). But that has not happened yet, and the private sector's role has been secondary to the city's in HI-TOADS redevelopment.

A central research question in this study has been an evaluation of how successful cities are with HI-TOADS reuse. I measure success by whether a property that was a HI-TOAD in the last decade is still dragging down neighboring property values in 2006. A close study of the HI-TOADS in Trenton shows that the city's efforts can be tied to the conversion of three HI-TOADS into "non-HI-TOADS," that is, that no longer meet the strict HI-TOAD site definition. Not to say that these three sites are all gems, but the interview and housing evidence demonstrate that they no longer can be seen as tumors or blights on their neighborhoods. At Crane, Roebling, and Magic Marker, public policies worked to eliminate the worst negative externalities. My statistical analysis of census data showed that the Crane and Roebling neighborhood did not perform very differently from the city as a whole during the periods before redevelopment or after. Conversely, the demolition of the Magic Marker structure in 1998 was accompanied by a strong performance in housing values in that census tract between 1990 and 2000, relative to the city as a whole. While this chapter does not prove that the Magic Marker's reuse caused improvements in neighborhood property values, the evidence suggests that the reuse did benefit the neighborhood.

While the quantitative evidence of success is very weak, my extensive interviews and field visits confirm that public sector action at Crane, Roebling, and Magic Marker was successful. Less clear is how to evaluate the city's efforts (or lack thereof) at the other five HI-TOADS. The city's policy formula varied site to site and began at different points. Efforts to address the Thropp and Assunpink Greenway are relatively new and the city's failure to date to redevelop these sites is understandable. In the next section of this chapter, I will describe the central policies and circumstances applied to address HI-TOADS and identify the key lessons from Trenton.

Key Themes in Trenton HI-TOADS Redevelopment

I present below five key themes developed as a result of a detailed analysis of the dimensions of HI-TOADS reuse planning and public policy in Trenton. The five themes correspond loosely with the five research questions that have driven much of the case-study work:

1. Did the city's policies fall into one or more of the broad categories identified in the literature review (chapter 2): economic development, community empowerment, or healthy environments;
2. How were HI-TOADS incorporated into the city's planning efforts?
3. Why did the city employ the strategies it employed?
4. How did the city manage the challenge of historic preservation with respect to the reuse of HI-TOADS?
5. What is the nature of NGO involvement in HI-TOADS?

In my effort to identify broad policy responses to HI-TOADS, I did not expect a city to be so organized and consistent as to have a single set of policies applied to all its sites. In Trenton, there is little consistency in the application of city policies. Economic development is used at some sites, community empowerment at others, and healthy environments at another. This varied application may reflect several factors. First, the unique needs and circumstances surrounding each site may merit a different type of city response. For example, sites with high market value due to location and infrastructure were targeted with economic development policies (Crane and Roebling). Community empowerment and healthy environment policies were applied to a site where there was high community concerns about environmental contamination (Magic Marker). And a site that had been subjected to unique flooding problems and was seen as no longer suitable for intense development was targeted with healthy environment policies (Assunpink Greenway). The city appeared to use location, market, environmental, and natural resource data to develop a rational basis for the application of various sets of policies and planning.

Early in Alan Mallach's tenure at the city, an effort was made to break the city into twenty-five zones for the purposes of advancing land-use planning. In explaining how the city operated without a current master plan, one former official said "[Trenton's] overall Master Plan was that we developed redevelopment plans for each neighborhood. We had a detailed plan for [many HI-TOADS], others it was informal, more opportunistic." Those plans were codified in Trenton's 1999 Land Use Plan. As explained in the Land Use Plan: "there are many areas in the city in need of change and improvement. Each of these areas is different and requires a different approach and a different series of planning and development strategies" (13). The plan focuses on "different visions for, and the redevelopment and land use issues of, twenty-five selected geographical sites or 'special planning areas'" (2).

For many of these special planning areas, the city developed Redevelopment Plans, in accord with the New Jersey Redevelopment Act. The adoption of a Redevelopment Plan under New Jersey state law provides a municipality broad authority to designate land uses, work directly with a private developer, and acquire property under condemnation. One of Mr. Mallach's accomplishments that was most respected among the officials I interviewed was how he was able to expand the interpretation of blight to include brownfields in crafting Redevelopment Plans. By doing so, Magic Marker, Crane, and Roebling all were placed successfully within Redevelopment Areas and, as such, the city had wide latitude to intervene as it saw necessary.

A state official was pleased with the zonal approach: The "city uses Redevelopment Plans for areas of the city, then attacks the HI-TOADS . . . then identifies obstacles in those areas." The special planning areas and Redevelopment Areas meld quite well with NJDEP's own area-wide solution, the BDA. As discussed earlier, the BDA's have helped to focus state resources on neighborhoods with HI-TOADS, designating one for the Assunpink Greenway and one for the Magic Marker area.

In sum, these area-wide efforts have helped to channel city and state attention to the neighborhood conditions around HI-TOADS. They are good examples of how the city was engaged in formal government planning, including goals and means setting through a public process and ultimately codified.

The BDA program is relatively new, so its effectiveness cannot be evaluated at this time. But Trenton's special planning areas and Redevelopment Areas did work effectively in spatially organizing land-use challenges in the city. Those efforts appeared to play a crucial role in success at Roebling, Magic Marker, and the Crane site.

. . .

Throughout the Trenton case study, one theme stood out as unique. In studying planning documents, in visiting sites, and in the interviews, I found that the City of Trenton did something that is rare in most depressed urban areas. For much of the last decade, the City of Trenton purposefully planned for its neighborhoods and followed through on those plans. Under Mayor Palmer and Mr. Mallach's leadership, the city looked closely at those 25 neighborhoods in need of special planning and generated a land-use vision for each. As one former city official put it, "Alan thought this stuff through, he wrote a Land Use Plan for the city."

The development of a vision for each special planning area was based on a careful consideration of planning and market condition, with remediation issues taking a back seat. The city sought to define visions based on good planning, which consisted of some blend of what community residents wanted at the site, separation of industrial and residential uses, catalytic properties of certain types of development (such as market-rate housing or parks), transportation considerations, historic preservation considerations, and public access considerations (for important public resources like rivers).

The approach was explained well by one former city official:

> The fundamental policy is that brownfields sites are opportunities for neighborhood revitalization. What the most appropriate use is in terms of revitalization of a neighborhood. Not to be determined by remediation. If site should be housing, then that is what is should be. If costs, time involved are too high, then maybe you have to shift gears . . . You've got to function within the constraints of the market, you've got to find something that can carry itself financially over time.

A current city official agreed. "Reuse was fueled by planning decisions instead of paths of least resistance." At a site like Magic Marker, the visioning process never contemplated a new factory at the site. "Nobody wanted Magic Marker to come back" remarked a representative of a community organization. She continued, "our intent is to re-imagine the city . . . the number of jobs is completely depleted . . . it's been cheaper and easier to build what was once built here, elsewhere." By combining a planning process with market analysis (with little consideration for environmental costs) the city has broken away from the past and chartered a new course for the future. Not only is this rare in depressed American cities, even more rare is that fact that Trenton has followed through on its plans and its visions and its implementing many of them, including actively addressing many HI-TOADS.

Of all, the most bold long-term vision is that for the Assunpink Greenway. While implementing the vision for a greenway is "not a priority citywide . . . [the city acts when] a source of funds comes up, or when developers or community groups are pushing" according to a former official. But the city is committed to its vision. "[The] city has come up with a great vision and has begun to implement it." A local community leader remarked: "Taking a floodplain and re-imagining what the city can be . . . I give them a lot of credit."

One former official fears that such a commitment to long-term planning is today being compromised in Trenton:

> Current activity is more reactive, if a developer comes in, they try to accommodate . . . It's completely night and day, one is more knee-jerk, a developer comes to you, someone who was trying to establish a policy . . . look to establish a long-term vision of how you want the city to develop, to prevent brownfields from occurring. [The knee-jerk response] is disastrous, [Mallach] had a five- to ten-year vision for neighborhoods . . . Staff is still following Alan's policies, but that all gets derailed as soon as a developer comes in and they try to accommodate them.

Today, Trenton is more famous for its past than for its present or future. Trenton's Revolutionary War significance, its proud contribution to the industrial revolution, and its stature as a major manufacturer in the early twentieth century together make it a great historical city. But a romantic past does not *necessarily* pay the bills. With much of the city's downtown comprised of city, county, state, and federal properties that do not pay property tax, the city has tremendous fiscal challenges. For centuries, that tax base was supported by a vast industrial complex that at one point hired 50,000 workers and today hires fewer than 3,000 workers (Andrews 1999).

In Trenton, officials appear to have accepted that shrinkage of manufacturing and are poised to try to remake the city by implementing the master plan. One member of the BEST Council explained that many of the HI-TOADS initiative stemmed from "recognition that the city was not going to go back to being an industrial center like the way Roebling was." The sentiment was further expressed in the city's 1999 Land Use Plan: "Trenton may never again be the industrial giant it once was, but there are important niches that the Trenton economy can successfully fill in both the industrial and the entertainment sectors" (10). City planners believe they have found a way to remake the city without destroying all that is special about its past: "As a central element of its vision of the future, Trenton must restore and preserve its historic character even as it remakes itself into a modern city with cultural and entertainment activities appropriate for the 21st century" (City of Trenton 1999, 8).

One of the prime mechanisms for achieving that vision is the conversion of several historically industrial zones of the city to other uses through zoning and designation of Redevelopment Areas. In the Land Use Plan, the city sets out to "convert land use patterns from the historical, but obsolete, industrial uses."

One community member saw the conversion of "obsolete, industrial uses" to residential as being essential for getting HI-TOADS back into productive reuse. "It's easier to do a residential reuse because there is always residential demand . . . trying to attract industry is a hard thing . . . that's the HI-TOADS challenge." Because many of these industrial structures no longer meet the needs of modern industry, they are no longer sought after for such uses in the private market. For example, the Crane project was successful in that the site is no longer a negative draw on the surrounding neighborhood. But this success occurred only after the infusion of vast public resources. However, it is these industrial uses that truly contribute to fiscal and social stability through property tax payments and as employment generators. While a residential use might sufficiently convert a site from being a HI-TOAD site, thus limiting its negative reach into the neighborhood, residential use may not always have the same larger, citywide benefits that a light industrial or mixed-use development might.

So when confronted with a tangible piece of the city's past, what do Trenton officials do? In most cases, they try to preserve historic structures, but in only a few cases are they successful. A planner described the city's efforts at the Crane site: "We were trying to save some of those buildings; it didn't work." A former city official described the city's historic preservation efforts at Roebling with contempt, "At Roebling, they did demolition first, then they went to SHPO [State Historic Preservation Officer]"—federal law requires that the SHPO be consulted first when historic resources are involved and federal dollars are funding the project.

The City of Trenton is trying desperately to capitalize on its HI-TOADS sites by making them part of a larger vision of a new, modern city. As one community leader explained when discussing the Magic Marker project, "Nobody wanted Magic Marker to come back," there is little nostalgia for the old industry that occupied many of the city's HI-TOADS. The vision for the future is primarily to accommodate new, perhaps cleaner, industrial uses to the city. When so much of the city's industrial lands are laying fallow, the easy answer is to reoccupy those sites with new industrial uses. However, those buildings' obsolescence for twenty-first-century industrial operations makes such a move quite challenging. Therefore, the city's strategy has been to demolish these kinds of structures to make way for new, property-tax-generating uses. Preservation has been accomplished only on a few isolated sites.

The implications of such a strategy are profound. By setting the city's past aside so cavalierly, Trenton is making strides toward establishing itself as a place of importance today and in the future, not just in the past. The costs of such a strategy are largely unknown.

In many ways, the NJDEP's involvement of Isles in the Magic Marker site made for a very successful pilot project. In other ways, that involvement typifies what is wrong with HI-TOADS redevelopment in Trenton because it was so unusual. Isles facilitated community involvement in the site, educated residents about the hazards at the site, organized the residents politically, and helped to generate a grassroots vision for the reuse of the site. While those efforts are commendable, the presence of community organizations in the reuse of the other seven HI-TOADS in Trenton was virtually nonexistent.

The initial idea for a community-driven clean-up was experimental in nature. NJDEP's then-director of research, Judy Shaw, asked whether a community-driven process would be faster and yield better benefits for residents than the normal state-, city-, or developer-driven clean-up process. By most accounts, the idea was a great one and that small grant back in 1993 was the catalyst for a true grassroots-driven process that resulted in the demolishment of the derelict Magic Marker structure and initiated the environmental remediation work that today is nearing completion.

Isles' work revolved around four key functions: facilitation, educating, organizing, and visioning. First, they worked with local residents to identify the most severe sites in the city, essentially the HI-TOADS. "These were the biggest and of the worst concern to residents who lived nearby," explained a former official.

A representative of a community organization said what Isles did "helped to educate the community around the Magic Marker property. It wasn't just another civic exercise, the city really took action . . . you had a city that was responsive." Isles "helped facilitate a process." Another community representative explained that Isles "started to organize the community to tell the mayor and the state that these are important issues." She continued, "the political process is part of the clean-up process—it's not a bunch of scientists making decisions. Scarce resources, someone who has to decide where the money goes and how to get on someone else's radar."

Isles also played an important role in visioning and planning for the future use of the Magic Marker site. "What [had] held back Magic Marker was that sense of what is should be. With Isles' help they finally did" develop a path forward to develop such a reuse plan, reported a BEST member. Another community member reiterated the point: "[The] final land use [plan] is very reflective of what residents want—low-rise housing and some open-space improvements."

However important the Isles story is at Magic Marker, the larger story is how little role community groups have played in the other seven HI-TOADS in Trenton. The success at Magic Marker has garnered the city awards and recognition for community involvement from both EPA and HUD: "The bureaucracy rests on its beneficent reputation and relies heavily on Isles to fulfill its mandate to involve the community. But it does so at the expense of other community organizations, which have had less success working with the city on brownfields issues" (Gardner 2001, 151).

Beyond Magic Marker, community involvement in HI-TOADS has been virtually nonexistent in Trenton. A former official described the planning effort for the Assunpink Greenway by borrowing from the language of the modernist planners. "We can do something really big here," yet conceded, "community groups did not play a significant role" in the development of the plan for the Greenway. The expression "we can something really big here" is key because the "we" does not include community groups or residents.[7] While there was some important early involvement of community groups at the Roebling Complex, that group disbanded more than a decade ago and the site remains nearly half vacant.

The city developed its Land Use Plan through an extensive public participation process. The ideas and visions for what should become of the city's neighborhoods were not generated directly from an authoritative leader but rather through an iterative process with a multitude of constituents. However, when it came to developing detailed plans and executing those plans, community groups and residents largely have been excluded (with the exception of Magic Marker).

The City of Trenton is not afraid of HI-TOADS. Efforts over the last decade especially have indicated not only a willingness, but an enthusiasm on the part of public officials at the local, county, state, and federal level to take on some of the more complicated and challenging abandoned sites in the city. Through a policy of acquisition, control, seeking external funding, and seeking development partners, the city has been heralded by both HUD and EPA as an exemplar on how to do brownfields redevelopment. With respect to HI-TOADS, those resources have yielded new jobs, new property tax revenues, and a cleaner local environment. Yet, despite all of the attention, six of the city's eight HI-TOADS remain idle. Public officials decry the lack of private investment and the scarcity of public resources. During the second stage of this research (when I conducted telephone interviews with local officials throughout the country), I learned about how historic preservation was used to promote HI-TOADS reuse, I learned about how the interim use of property was a powerful tool, and I learned about the effectiveness of comprehensive planning.

The city's lack of genuine commitment to historic preservation was important for how it executed its redevelopment. But, it may be that Trenton has something to learn from the experiences of other cities that built historic preservation more integrally into their HI-TOADS reuse policies.

Throughout the case-study research, I did not learn of any concerted effort on the part of Trenton to put abandoned HI-TOADS to some sort of interim use. Evidence from the telephone interviews suggests that a demonstrated presence on HI-TOADS ameliorates their negative externalities.

And lastly, while Trenton's long-term planning approach was essential to its success, the absence of a comprehensive, unitary plan for the city appears to be a central flaw. The designation of scores of special planning areas was valuable, but the failure of the city to look at the big-picture forces affecting its social, economic, fiscal, environmental, and land-use conditions inhibited city officials from being able to rethink and remake Trenton. HI-TOADS can serve as the heart of such efforts. For the City of Trenton, that chance has not yet been embraced.

[6]

Slag Heaps, Steel Mills, and Sears

PITTSBURGH, PENNSYLVANIA

With the story of how New Jersey's capital city addresses its HI-TOADS complete, I now turn to a large city on the edge of the Midwest. To many, Pittsburgh is a symbol of successful urban revitalization. From a depressed old steel town in the 1970s to top city for quality of life in the twenty-first century, Pittsburgh has experienced a remarkable rebirth (Economist Intelligence Unit 2006; Savageau 2007). Because the city is more than 99 percent built-out, new development in Pittsburgh is almost always on brownfields.

Studies of brownfields redevelopment in Pittsburgh have uniformly celebrated the city's achievements and held it up as a model for other depressed Rust Belt cities (Davis 2002; U.S. Environmental Protection Agency 2006b; Stikkers and Tarr 1999). As discussed in chapter 2, greater insight into neighborhood change can be realized by looking at urban redevelopment through the HI-TOADS lens. For all the acclaim Pittsburgh has received for its renaissance, how successful has it been in reusing HI-TOADS?

As I did in the Trenton case study, I will begin by reviewing the key study questions for Pittsburgh. Next, I will provide a brief history of Pittsburgh, with an emphasis on current political and economic conditions. Then, I will profile four HI-TOADS in Pittsburgh, detailing the stories of their reuse or lack thereof. In the following section, I will describe how various players at the governmental and nongovernmental levels have addressed HI-TOADS in Pittsburgh. Lastly, I will present five key themes that derive from the case-study analysis probing the dimensions of HI-TOADS reuse planning and public policy in Pittsburgh.

Key Research Questions for Pittsburgh

As in the Trenton case study, two broad questions are addressed here: (1) of those cities that acknowledge HI-TOADS as a problem, what policies do they use to address them? and (2) how successful are those policies? Within the first question, I am attempting to categorize each cities' efforts into one of three classes of policies: economic development, community empowerment, or healthy environment. In my interviews with the two Pittsburgh officials in Stage 2 of the research, three additional questions emerged for exploration in this case study.

First, it appears that community groups and foundations play a central role in HI-TOADS reuse in Pittsburgh, yet the city did not have any formal avenues for their involvement. In this case study, what role do community groups play in HI-TOADS reuse in Pittsburgh and how does the city formally or informally support or discourage that?

Second, Pittsburgh does not have an updated comprehensive plan. Its City Planning Department is not heavily involved in HI-TOADS reuse. City officials at the Urban Renewal Authority (URA) of Pittsburgh are focused primarily on utilizing HI-TOADS for economic development purposes. This economic development orientation automatically preempts a wider, community-based planning process (like the kind that occurs through the development of a comprehensive plan). Given these conditions, what is the role of planning in determining future uses of HI-TOADS?

And third, where Trenton's story is one of a thoroughly depressed city with little private investment interest, Pittsburgh is a vital economic center for its region and has continued (despite its difficulties in the last several decades) to attract external investment and new real estate development (Miller 2005). Despite such a relative position of strength, officials reported six HI-TOADS in the city. How does Pittsburgh's warmer development climate play into city policies for HI-TOADS reuse?

At the famous point where both the Monongahela and Allegheny rivers form at the Ohio River, Fort Pitt was established in the early seventeenth century. The fortification played a critical role in the French and Indian War of the 1760s and soon a small city grew around the fort (Tarr 2003). The City of Pittsburgh was incorporated in 1794.

As in Trenton, the industrial revolution made its mark on the Pittsburgh landscape. While all forms of industry took hold in the booming nineteenth-century city, steel and iron production dominated. Pittsburgh was "the industrial center of the universe," according to one observer (King 2006, 1). Consuming wide swaths of land, almost always along the city's rivers, the busy employment centers attracted immigrants from all over the world (City of Pittsburgh 2001). But as industrial centers in other regions grew and prospered, the Pittsburgh area's competitive advantage weakened. By the 1940s, Pittsburgh's once-great reputation had turned to mud: "In 1943, the *Chicago Tribune* dismissed Pittsburgh, saying it wasn't a major city. In 1944, the *Wall Street Journal* rated Pittsburgh as a 'class D' city, with little hope for recovery. Smoke clogged the air, sewage roiled the waters and rats combed the fringes of Downtown" (Fitzpatrick 2000a). Experimentation with urban renewal in the 1950s and 1960s was largely unsuccessful. Civic and governmental organizations worked to stop the flow of people and jobs out of the city. These efforts failed and the city's population fell over half from 676,000 in 1950 to 335,000 in 2000 (U.S. Census 2000).

In 1994, Mayor Tom Murphy was sworn in as Pittsburgh's 57th mayor. Whether it was pure luck or whether he deserves the credit, around that time, Pittsburgh's fortunes began to change. Over the next decade, Pittsburgh experienced a dramatic turn-around. "We lost 146,000 manufacturing jobs, then replaced them with 168,000 education and medically related positions," said Michael Edwards of the Pittsburgh Downtown Partnership (quoted in O'Toole 2005, 7). Less clear is how the city's rebirth will affect its HI-TOADS. With so much of the city built-out, HI-TOADS are some of the few remaining places for this new growth to go.

HI-TOADS in Pittsburgh

Five HI-TOADS were identified in Pittsburgh: South Side Works, Nine Mile Run, the Hazelwood LTV Site, the Heppenstall Site, and the Sears Site (see table 6.1 and figure 6.1). In this section of the chapter, I will describe four of them in detail, provide background on reuse and redevelopment efforts, and present the results of statistical analysis of census data for each site. I will then synthesize the data analysis for all five HI-TOADS and present a summary of what I found with respect to HI-TOADS redevelopment, socio-economic changes, and property values in Pittsburgh (see table 6.2).

Table 6.1
Summary Information for HI-TOADS in Pittsburgh

Site name	Size	Location	Historical use	Proposed use	Status of redevelopment efforts (as of 2006)
South Side Works	123 acres	Between Carson Street and Monongahela River in South Side neighborhood	Steel mill	Mixed use	The Urban Redevelopment Authority (URA) of Pittsburgh is nearing completion of reuse of the entire site for retail, housing, office, park, and educational uses.
Nine Mile Run	250 acres	Between Interstate 376 and the Monongahela River near Frick Park	Steel slag dump	Residential	URA developed the site with 713 homes and 114 acres of park land.
Hazelwood LTV Site	200± acres	Between 2nd Ave. and Monongahela River in Hazlewood neighborhood	Steel mill	Mixed use	A consortium of nonprofit organizations and the Regional Industrial Development Corporation of Southwestern Pennsylvania purchased the property in 2002 and are in the midst of developing reuse plans.
Heppenstall Site	14 acres	At Hatfield Street between 43rd Street and 48th Street in the Lawrenceville neighborhood	Knife manufacturing	Flex office-industrial	The Regional Industrial Development Corporation of Southwestern Pennsylvania purchased the property and is developing reuse plans.
Sears Site	6± acres	At intersection of North Highland Avenue and Rodman Street in East Liberty neighborhood	Retail	Retail	With major city involvement, Home Depot built a new store at the site.

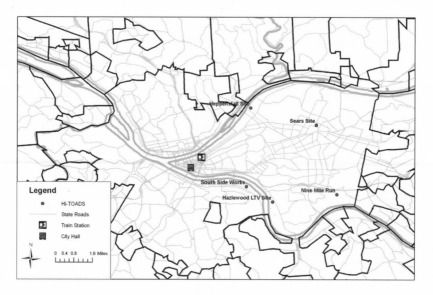

FIGURE 6.1. HI-TOADS in Pittsburgh, Pennsylvania.

South Side Works

Beginning in 1853, the South Side Works was a fixture on the south side of the Monongahela River for over one hundred years. Comprised of 123 acres, the site was known to have employed over 43,000 workers at its peak (Hartkopf 2005). In 1986, when the facility was closed, reverberations were felt throughout the neighborhood. David Lewis, professor of urban design at Carnegie Mellon University, commented at the time that the closure was "totally undermining the economic base of the community. Shops and pubs closed, real estate values tumbled leading the entire area into a period of decay" (as quoted in Hartkopf 2005, 2). Six years after shutdown, the site owners, the LTV Corporation, reached out to the local community development corporation to plan for a new use of the site.

The South Side Local Development Corporation (SSLDC) already had garnered itself a strong reputation in the neighborhood, as both an advocate in City Hall and as an active developer. This was partially due to its enlistment of Rebecca Flora as its executive director. Ms. Flora, a member of the American Institute of Certified Planners, had been the Urban Redevelopment Authority (URA) of Pittsburgh's project manager for several large brownfields redevelopment projects.

Table 6.2

Property Value and Demographic Data for HI-TOADS and the City of Pittsburgh, 1980 to 2000

	1980	1990	2000	Percent change		
				1980–1990	1990–2000	1980–2000
City of Pittsburgh						
Mean housing value ($1,000s)	37.8	22.3	43.8	-41	96	16
Median gross rent ($)	185	372	500	101	34	170
Number of occupied housing units	166,067	153,452	143,739	-8	-6	-13
Average household income ($1,000s)	13.4	23.0	28.6	72	24	113
Population	423938	369829	334563	-13	-10	-21
Percent African American	24	26	27	8	5	13
Percent Hispanic	1	1	1	0	30	30
Heppenstall						
Mean housing value ($1,000s)	8.0	13.0	16.8	63	29	110
Median gross rent ($)	181	320	466	77	46	157
Number of occupied housing units	1142	1010	959	-12	-5	-16
Average household income ($1,000s)	14.1	28.6	38.7	103	35	174
Population	2753	2347	1881	-15	-20	-32
Percent African American	0.0	0.0	4.0	—	—	—
Percent Hispanic	0.0	0.0	0.0	—	—	—

Table 6.2 (continued)

	1980	1990	2000	Percent change		
				1980–1990	1990–2000	1980–2000
Sears						
Mean housing value ($1,000s)	2.3	3.7	4.9	61	32	113
Median gross rent ($)	445	773	420	74	-46	-6
Number of occupied housing units	2800	2695	2124	-4	-21	-24
Average household income ($1,000s)	12.0	17.9	29.7	49	66	148
Population	5051	4814	3791	-5	-21	-25
Percent African American	62	65	79	5	22	27
Percent Hispanic	2	1	1	-50	0	-50
*9 Mile Run**						
Mean housing value ($1,000s)	23.5	32.2	55.4	37	72	136
Median gross rent ($)	413	458	602	11	31	46
Number of occupied housing units	3822	3587	3679	-6	3	-4
Average household income ($1,000s)	21.9	72.7	54.9	232	-24	151
Population	9340	8011	7896	-14	-1	-15
Percent African American	2	4	3	100	-25	50
Percent Hispanic	0	1	2	—	100	—

Hazelwood LTV Site

Mean housing value ($1,000s)	12.9	16.7	31.2	29	87	142
Median gross rent ($)	358	702	397	96	-43	11
Number of occupied housing units	1091	1026	849	-6	-17	-22
Average household income ($1,000s)	15.5	23.2	31.6	50	36	104
Population	2953	2495	1968	-16	-21	-33
Percent African American	18	22	20	22	-9	11
Percent Hispanic	1	1	0	0	-100	-100

South Side Works

Mean housing value ($1,000s)	5.4	9.9	17.6	83	78	226
Median gross rent ($)	124	326	483	163	48	290
Number of occupied housing units	1237	1256	1160	2	-8	-6
Average household income ($1,000s)	12.3	18.8	36.0	53	91	193
Population	2643	2339	2111	-12	-10	-20
Percent African American	4	3	4	-25	33	0
Percent Hispanic	1	0	1	-100	-100	0

*Nine Mile Run straddles two census tracts; all other sites are enclosed within a single tract. Data presented here for Nine Mile Run includes both tracts.

By the time that SSLDC was commissioned to develop a community-based reuse plan for the South Side Works site, according to Ms. Flora, it already had made great strides in turning the neighborhood's business district around and upgrading a significant portion of area housing stock. The resulting neighborhood-wide plan looked to the South Side Works site as a catalyst for much of the envisioned revitalization. Ms. Flora was the main author of the plan, which was an important touchstone for the subsequent efforts to reuse the site.

Shortly after the adoption of the plan, Mayor Murphy directed the URA to acquire the site for $9.3 million with the intention of managing the redevelopment. One former official explained that by controlling the redevelopment, the administration believed that it could direct the reuse to something other than "Butler buildings and at-grade parking."[1] South Side Works "was a place to create value." One official said that there was a "need to protect it from the bottomfeeders."

After acquiring the property, the city issued a request for proposals. They received interest first from Hospitality Franchise Systems, who was interested in using the property for riverboat gambling. After working

FIGURE 6.2. South Side Works: Gentrified Carson Street in the foreground, with the newly developed South Side Works complex in the background.

FIGURE 6.3. South Side Works: Looking across Hot Metal Street in the heart of the South Side Works development, with downtown Pittsburgh in the background.

with HFS for several years, the URA then partnered with the Soffer Organization. The URA executed its own reuse plan in 1997, The South Side Works Preliminary Land Development Plan. Community reaction to that plan was mixed, so with funding from the Heinz Endowments and led by the Green Building Alliance, a third study was initiated to examine gaps between the original 1992 reuse plan, the 1997 URA Plan, and the work at the site, to date.[2] The result was a better integration of the original community vision into the final stages of the redevelopment. While work is continuing today, the bulk of the South Side Works site has been reused (see figures 6.2 and 6.3). According to officials interviewed, the site was no longer a HI-TOAD after the initial redevelopment occurred in 1998. Instead of the site bringing down neighborhood property values, as it did before 1998, the new development projects began to have a positive external effect on the surrounding neighborhood. As evidenced by the observations of one local community leader: "most of South Side's upswing happened after South Side Works."

A look at census data validates those conclusions (see table 6.2). From 1990 to 2000, both median gross rent and mean housing values rose at a rate comparable to the city as a whole. A growth of 78 percent in mean housing value still puts the area at significantly lower levels than the city as a whole ($17,600 per housing unit as compared to $43,800 per housing unit in all of Pittsburgh). But housing values have grown dramatically since the 2000 census. An official with the URA reported that in 2006, "it's tough to find a house down here for less than $300,000." The rapid escalation in housing prices over the last six years can be attributed partially to the city's efforts at South Side Works, but also can be attributed to the wider upsurge in values in the gentrifying neighborhoods to the north and west of the site.

The South Side Works project was initiated through a community empowerment process, yet it was largely outside the city's involvement. That is, the city did not directly support a community empowerment process. Rather, once Mayor Murphy decided to acquire the site, the city applied its full range of aggresive policies to ensure an economic development orientation for the project.

Nine Mile Run

Wedged between the Mononghela River and Interstate 376, Nine Mile Run was a 250-acre dumping ground for slag, a major waste product of the iron- and steel-making processes. For fifty years, from 1922 to 1972, slag from throughout the region was poured into what had been a pristine valley (McElwaine 2003). In the end, 17 million cubic yards of slag filled the valley, at points piled as high as 120 feet (McElwaine 2003).

When dumping ended, a wide range of actors, including the city and civic organizations, began to debate possible reuse plans. Eminent historian Joel Tarr explained the result: "The complexities of site development, however, were large, and funding and developers were hard to find. Issues relating to traffic and neighborhood opposition were especially significant, and none of these plans became a reality" (Tarr 2002, 536).

The site remained idle until the early 1990s, when a coalition of nongovernmental organizations decided to embark on a master development plan for the Squirrel Hill neighborhood, one of Pittsburgh's most stable. The coalition raised $70,000 (of which $20,000 came from the city) and hired Urban Design Associates, a local planning and design firm, to devise the plan. Through an extensive public planning process, the nearby Nine Mile Run site was identified as underutilized and became the subject of closer study. According to one of the community leaders involved at the time, the planning team "started to wonder if there was some way to use it for productive purposes. That concept of using the slag dump was a component of the neighborhood plan."

The neighborhood plan was first put to action when several local Jewish organizations sought to expand their operations with a new school. "The idea of Nine Mile Run got some attention" during the debate over the new school, explained a community leader. He continued, "One of the elected officials said if we can change the slag heap into something desirable it would mitigate the impacts" of the new school project. Quickly, other political officials agreed and a redevelopment of the Nine Mile Run site was universally seen as a good idea.

This feverish activity culminated when the Pittsburgh Urban Redevelopment Authority (URA), under the direction of Mayor Murphy, purchased the site in 1995 (Tarr 2002). Mayor Murphy is an avid runner and bicyclist. As a resident of the neighborhood near Nine Mile Run, he regularly would run near the site. For Mayor Murphy, the choice to put city resources into the site was easy. He recognized that nobody else was going to take on the project with all of its unknowns. A former city official summed up the city's policy at the site: "You need money to be able to get in the game, control the land, then go out and find the best partners, then reasonably be able to share the risk."

As a result of the city's acquisition and subsequent partnering with a private development company, Summerset Land Development Associates, a new housing development of 713 homes and 114 acres of park land was created (see figure 6.4). The city paid $3.8 million for the site and invested millions more for environmental investigations and an economic development package to the developer (Tarr 2003).[3] But the investment did successfully eliminate a HI-TOAD site and in its place has attracted significant new tax revenues.

While planning for what became Summerset at Frick Park began in the early 1990s, site grading only began in 1999. Analysis of census data shows that in both 1990 and 2000, the census tracts that host Nine Mile Run are collectively wealthier (higher household income) and have higher property values and rents than the whole of Pittsburgh (see table 6.2 for details). While grading occurred in 1999, significant activity involving environmental investigation and remediation occurred prior to that date. In studying

FIGURE 6.4. Nine Mile Run: View from the top of the former slag pile down to Phase I of Summerset toward Frick Park. Photograph by Marc Knezevich.

the census data, the key time frame of concern is from 1990 to 2000. During that period, mean housing values rose 73 percent, while they rose 96 percent citywide. Also during that time, median gross rent rose 31 percent and 34 percent citywide. These changes appear to be evidence that the neighborhoods surrounding Nine Mile Run have almost kept par with citywide housing performance during the time of the preliminary site development. But in 2006, there is no question about the result of the development. One current official called the project "tremendously successful . . . a slag dump as the hottest neighborhood in the City of Pittsburgh? Shows you it can be done if you have an aggressive agenda."

The census data also helps paint a picture of the Nine Mile Run neighborhoods as largely white and relatively affluent, strikingly so in contrast to the city as a whole. In the interviews, community leaders raised key environmental justice questions about why public resources were focused on redeveloping a moribund site in this neighborhood and whether other pressing needs were neglected in poor, largely nonwhite neighborhoods.

Like South Side Works, the original neighborhood plan that kicked off a vision for residential development of Nine Mile Run was completed by community-based organizations. However, at Nine Mile Run, the city played an important role in sponsoring and supporting that neighborhood effort and, in time, executing the community-derived vision through the reuse of the site. But the project itself is a textbook example of economic development: When complete, it will provide more than 700 new housing units on just over 200 acres in a central urban location, with over 100 acres of new park land. In effect, this project was a combined economic development and community empowerment initiative. Elements of both sets of policies were applied simultaneously to the site.

Hazelwood LTV Site

While giving me a tour of South Side Works, on the south shore of the Monongahela River, the URA's spokesperson, Julie Desyn, pointed across the river to the Hazelwood LTV site (see figure 6.5). "That's one we're not involved with at this point." Sitting on the northern shore of the Monongahela River, this 200-plus-acre former LTV coke works plant is almost a mirror image of what the South Side Works site looked like just ten years ago (see figure 6.6). Today, they do not bear any resemblance to each other. The LTV site is comprised of scores of abandoned structures, including sheds, offices, and warehouses. Only a couple hundred feet deep, the site stretches for over a mile, butting up against the Hazelwood neighborhood the whole way.

Once one of Pittsburgh's great commercial centers, Hazelwood has seen better days. High crime levels and high unemployment have hit the neighborhood hard and hopes for a bright tomorrow are tied closely into a future reuse of the shuttered LTV plant (City of Pittsburgh 2005). The LTV coke works plant shut down in 1998 and since then, it has been a nuisance for the neighborhood. One local community activist expressed his concerns, "folks can get in, local residents sneak through the fence."

A county official explained the larger context for redevelopment at the LTV site: "One of the growth areas in southwest Pennsylvania is Oakland, it's bursting at the seams. The Hazelwood site is tailormade to accommodate that growth . . . The problem is the highway, it will slice up the site." The highway Mr. Davin is referring to is the final extension of the Mon Fayette Expressway, slated to cut through the Hazelwood neighborhood and

FIGURE 6.5. Hazelwood LTV Site: View from South Side Works, across the Monongahela River.

effectively serve as a barrier between the LTV site and the remainder of the neighborhood (City of Pittsburgh 2005). But notwithstanding the planned highway, officials are confident that the spillover of office space demand from neighboring Oakland is well suited for the LTV site.

While the city was investing heavily in Nine Mile Run, South Side Works, and other brownfields, when the opportunity to acquire the LTV site arose, the city passed. "The city most likely would've bought that site if we hadn't already spent all our money" on other brownfields, remarked a city official. Another official contradicted him by explaining that "we felt the property had a negative value" (due to environmental contamination) and thus opted not to take on the risks of owning and redeveloping the site.

What is known for certain is that, instead of acquiring the LTV site, the city invested in two neighborhood planning efforts, first the Master Development Planning in Hazelwood and Junction Hollow report in 2001 and then the Hazelwood Second Avenue Design Strategy in 2005. The 2001 study sought to explore "the kinds of economic activities that should take place on the LTV site," while the 2005 plan recommended "a flexible physical framework" to ensure that the LTV reuse is compatible with and supportive of the existing Hazelwood neighborhood.

In 2002, a consortium of Pittsburgh-based foundations, in collaboration with the Regional Industrial Development Corporation of Southwestern Pennsylvania (RIDC) formed a new organization by the name of AL-MONO, LLP, and purchased the LTV site (City of Pittsburgh 2005). A leader of a local community organization explained why ALMONO stepped forward: "As stewards of the region, they saw an opportunity to invest with a program-related investment, as opposed to grant making. It will redevelop a brownfields . . . they're expecting a payback." A city official surmised that the "foundations purchased the Hazelwood site to prevent the Waterfront group from getting their hands on it, and to some extent they bought it to keep it out of the hands of the city."[4] Others in the community viewed the ALMONO move as a way to keep the mayor and the URA out of the redevelopment. Regardless of why ALMONO took on the redevelopment, the fact is that the site remains a HI-TOAD today. ALMONO's planning continues and they anticipate adopting a master development plan in 2007.

Concerns about environmental justice raised in the Nine Mile Run profile take on a particular severity upon a close review of census data for the Hazelwood neighborhood. This relatively poor, heavily African-American neighborhood has suffered from continuing depopulation in recent decades.

FIGURE 6.6. Hazelwood LTV Site: View from Scotch Bottom neighborhood, with downtown Pittsburgh in the background.

The neighborhood lost one-third of its population from 1980 to 2000. During that same time, the neighborhood experienced widespread housing abandonment. The number of occupied housing units fell from 1,091 in 1980 to 849 in 2000, a drop of 22 percent. As can be expected, the neighborhood has had persistently low housing values, relative to the whole of Pittsburgh.

The site has not been redeveloped, there are no immediate prospects for new uses of the site, and uncertainty about the future clouds real estate investment for the entire neighborhood. Mean housing values from 1980 to 2000 have risen only by $18,300 per housing unit, to $31,200. This still puts the neighborhood's mean housing value well below the citywide value in 2000 of $43,800. Even worse performing are median gross rents for the neighborhood, which climbed only 11 percent from 1980 to 2000, where they grew 170 percent citywide during that period.

With respect to policy orientation, the city has adopted a community empowerment approach to the Hazelwood LTV site. By funding and providing technical support to community-based efforts to plan for the reuse of the site, the city empowered local people and organizations to conceive of a new use of the site. While some city officials may have sought a chance to own and control the Hazelwood LTV site, ultimately the city did not take that action.

Sears Site

According to the East Liberty Quarter Chamber of Commerce, downtown East Liberty was the third-largest business center in the Commonwealth of Pennsylvania, after Center City Philadelphia and Downtown Pittsburgh. But by the 1950s, the downtown area was in decline and became the target of a federally financed urban-renewal effort by the URA. The East Liberty project is a prime example of urban renewal gone bad. Citywide, urban renewal "swallowed more than 1,000 acres of land, razed more than 3,700 buildings, relocated more than 1,500 businesses and uprooted more than 5,000 families" (Fitzpatrick 2000a).

The East Liberty urban-renewal plan was a fanciful scheme whereby a ring road encircled the main business district, which was converted to a pedestrian mall. Today's critics uniformly agree that this reconfiguration did little to revitalize the distressed neighborhood. Some even argue that it hurt East Liberty (Fitzpatrick 2000b). "We now have intense land-use dysfunction" said a local community leader.

One result of the urban renewal was the construction of a new Sears department store. For almost three decades, the Sears operated profitably in the neighborhood. But, in the late 1980s, the roughly 6-acre store closed. It was seen by one community leader as "the nail in the coffin" for East Liberty. The site lay fallow for almost a decade. One community member commented that "it was becoming an eyesore, it was neglected, we had environmental concerns. The longer that property remained vacant, it would encourage crime." A former official remarked, "it was a poison . . . it was like a cancer" on the neighborhood.

But it was the East Liberty Chamber that is credited with applying pressure to the city and the URA to address this HI-TOAD site. "It was really the business people's interests that were driving that agenda," said a local community representative. In 1994, the URA acquired the site and identified Home Depot as a potential future user. At first, Home Depot executives expressed little interest in an urban location with poor highway access. But Mayor Murphy was committed to the Home Depot use. He called the mayor of Atlanta, where Home Depot has its corporate headquarters, and asked the mayor to tell him about the two founders of Home Depot, Bernard Marcus and Arthur Blank. Something Mayor Murphy learned was that Mr. Marcus was heavily involved with Jewish philanthropic causes. Mayor Murphy then contacted leaders of the Jewish community in Pittsburgh and they invited Mr. Marcus to Pittsburgh. Mr. Marcus gave a speech to the Young Presidents Organization and afterward Mayor Murphy took him on a tour of the East Liberty neighborhood and made the pitch for the Sears site. Within three years, ground was broken on a new Home Depot, one of the first in an urban setting without proximate highway access. Through a wide array of city incentives, grants, and environmental assessment assistance, Home Depot was able to make the move (see figures 6.7 and 6.8).

"Home Depot was the catalyst" for some of the private-sector-initiated redevelopment occurring in the East Liberty area, concludes one former city official. A community leader in another neighborhood of Pittsburgh said, "clearly within a quarter-mile, it actually did start stimulating other activity in East Liberty." Teresa Lindeman (2001), a reporter for the *Pittsburgh Post-Gazette,* agrees that "momentum is beginning to build in the community around [Home Depot]." But despite the excitement, one community representative conceded that the neighborhood "still has a long way to go."

FIGURE 6.7. Sears Site: View of new Home Depot store from Highland Avenue.

A review of census data supports a cautious optimism. Given that the construction began in 1997, the site essentially was converted from being a HI-TOAD site to no longer being one in 1997. Therefore, the timeframe of statistical analysis is 1990 to 2000. During that time, mean housing values in the Sears Site census tract grew by 32 percent, from $3,700 to $4,900. But despite that growth, the neighborhood's values were still just a tenth of the citywide mean value of $43,800 in 2000. Rents in the neighborhood were erratic during the period from 1980 to 2000 and spiked to $773 per month, a figure twice the citywide average in 1990. Therefore, not much can be concluded by comparing changes in rent. Despite the celebration over the Sears Site success, the surrounding census tract continued to slide in terms of population (21 percent loss) and number of occupied housing units (21 percent loss). Those changes may be attributed to a wider URA effort in East Liberty to "fix" the urban-renewal mistakes of the past with further housing demolition and displacement. The URA and the Pittsburgh Housing Authority demolished several public housing developments and free-standing substandard housing in recent years, which may account for some of the population and housing units declines.

The city's aggressive role in the Sears Site is exemplary of economic development planning. In close coordination with the Chamber of Commerce, the city effectively shut out community-based organizations and local residents and directed a reuse to maximize economic use of the site.

How Are HI-TOADS Addressed in Pittsburgh?

Numerous former and current city officials emphasized that the central tenet in the city's approach to HI-TOADS was to obtain control of the site. For the South Side Works and Nine Mile Run, that was the case. However, the city government did not acquire the other three HI-TOADS in the city, the Sears Site, the Hazelwood LTV Site, or the Heppenstall Site. That is, the city's role in HI-TOADS was limited primarily to efforts at South Side Works and Nine Mile Run. That said, the city did influence the Sears Site redevelopment and sought and failed to play a controlling role in the Hazelwood LTV Site.

Beginning by exerting direct control of HI-TOADS, the city, acting primarily through the URA, would then initiate environmental investigations

FIGURE 6.8. Sears Site: Distressed residential area (on the right) across the street from the Home Depot.

and remediation projects, seek a developer partner, and then jointly redevelop the site in close coordination with that partner. That model, as was described above, was implemented effectively at both South Side Works and Nine Mile Run.

The leadership and vision of Mayor Murphy was also an important aspect of understanding what it was that the city did. The mayor's activist boosterism at the Sears Site is proof of his uncommon commitment to getting HI-TOADS reused. Additionally, the mayor's vision for Pittsburgh put brownfields and HI-TOADS in the cross-hairs. One former official explained:

> One problem Pittsburgh had was a civic depression . . . "what are we going to do to get the steel mills back?" people asked. The mayor understood that [the steel mills] were not coming back. He began to *get* part of this civic vocabulary—find new ways for people to conceive of this property [the former steel mill sites] . . . He was a former community organizer, his father was a steelworker . . . He was a visionary, so dedicated, he cares for the city. He said "I will not preside over a shrinking city"—Pittsburghers have always settled for "it will do."

But for Mayor Murphy, "it will do" was not enough. His boldness and risk-taking helped to create opportunities where developers saw none. These unique leadership qualities translated into the reuse of three of the city's five HI-TOADS (over the last ten years).

The local nature of HI-TOADS has meant that in Pittsburgh, the city is the main public redevelopment player. Allegheny County, the State of Pennsylvania, and the federal government all play an important, yet minor role.

Allegheny County is comprised of 130 municipalities, with the City of Pittsburgh at its center. The county has experienced the same kind of decline and HI-TOADS proliferation seen in the city. It too, was at one time a major host to steel- and iron-making operations (in addition to mining) that occupied wide swaths of land. So when the county government looks to invest in redeveloping HI-TOADS, it devotes the bulk of its resources to either supporting city efforts or to investing in areas of the county outside the City of Pittsburgh.

County resources have been useful to the city, but one state official summed it up well: "pretty much the city does its thing and the county does its thing." The county's most significant foray into HI-TOADS redevelopment was just outside of Pittsburgh in the Town of Homestead. That massive reuse of a steel plant resulted in millions of square feet of new retail and office uses. But, within the City of Pittsburgh, the county leaves development to the URA. This is in contrast with Trenton, where the county competes with the city in development projects.

The State of Pennsylvania has also deferred to the URA as developer, but has helped with financing both South Side Works and Nine Mile Run. One official lauded the state's efforts: "The state has played a pretty big role . . . we have a number of programs that help . . . our major goal under the Rendell Administration is the revitalization of older, previously developed sites, brownfields redevelopment. Ninety different grant and loan programs . . . target older, previously abandoned sites." These ninety state programs include loans to private developers and grants to public/nonprofits through the Industrial Site Reuse Funds. These funds cover 20 percent of the cost of environmental assessment and up to 75 percent of actual remediation. A state official remarked that these program "help level the playing field for developers [in comparison with greenfield sites]." The official went on to say, "the URA did work directly with the state to take advantage of all of these sources of funding" for both the South Side Works and Nine Mile Run projects.

The federal government was also an important player in financing South Side Works and Nine Mile Run. Grants, technical assistance, and loans from the Environmental Protection Agency (EPA) and the Economic Development Administration (EDA) were pivotal for both projects. Of particular note was a $200,000 U.S. EPA Brownfields Pilot Grant awarded to the URA in the early 1990s, used to conduct preliminary environmental assessments at Nine Mile Run.

An active community group was associated with each of the five HI-TOADS in Pittsburgh. A nonprofit developer, the RIDC, has also been active in reuse activities at the Hazelwood LTV and Heppenstall sites. In addition, the city's philanthropy community plays an important role in financing planning efforts at HI-TOADS and directly heads the reuse of the Hazelwood LTV Site. Lastly, the city's real estate development community has been involved in HI-TOADS projects, both in partnership with the city and independently.

Over the last decade, when capacity to address HI-TOADS was low among Trenton's community groups, the reverse was true in Pittsburgh. The South Side Local Development Company, a well-financed, professionally staffed nonprofit, led early community efforts to plan for the reuse of

South Side Works and played a key role in shaping development reuse plans. The Squirrel Hill Urban Coalition, Swisshelm Park Civic Association, and other area nonprofits worked together to develop a neighborhood plan for their community that served as the catalyst for a citywide effort to redevelop Nine Mile Run. At the Hazelwood LTV Site, the Hazelwood Initiative nonprofit group has been an active force in representing residents and developing reuse visions and plans. Each of these is a strong organization with deep roots in the community and professional staffs.

These community groups have thrived in part due to financial contributions from Pittsburgh's philanthropic community, "For it's size, the city has an extremely philanthropic community" commented one former city official. The Carnegies, Mellons, and Heinzs, among others, are renowned for their vast philanthropic endeavors, particularly for investments in the Pittsburgh region (Lagemann 1992). Their seed money was essential in early planning at Nine Mile Run and their acquisition of the Hazelwood LTV Site was testament to their commitment to HI-TOADS reuse. One city official felt that ALMONO's acquisition of the Hazelwood site was done in part to ensure that redevelopment was beneficial for the neighborhood and "to some extent . . . to keep it out of the hands of the city." The ALMONO perspective was that the city would redevelop the LTV site without concern for community benefits.

Where private developers in Trenton only got involved in HI-TOADS with strong government support, private developers in Pittsburgh were more active in investing their money in HI-TOADS. In this case study, I found no examples of a developer financing a HI-TOAD site reuse in Pittsburgh independent of public subsidy. But the investment climate in Pittsburgh seems to be warmer than in Trenton, and with the right amount of subsidy, private investors were interested in HI-TOADS. Private developers in one case were engaged in a bidding war to develop HI-TOADS. The real estate market is vital in Pittsburgh and investors are willing to take on the risk of a HI-TOAD site for the potential financial returns of redevelopment.

The foundations, the URA, the RIDC, and the community groups each work for HI-TOADS reuse but often in an antagonistic and competitive manner. When I talked to city officials, current and past, they criticized the foundations and the RIDC. When I spoke to staff at the RIDC and community groups, they are quick to blame the URA for delays and inaction. But the end result in Pittsburgh is quite different from Trenton. In Trenton, few community groups have the capacity to address HI-TOADS and the philanthropy community is essentially absent from this arena. The Pittsburgh story is one of numerous, competing actors working to redevelop HI-TOADS. The volume and ferocity of these efforts has meant that Pittsburgh has a pretty good track record of getting these sites back into productive reuse.

Key Themes in Pittsburgh HI-TOADS Redevelopment

Despite Pittsburgh's legacy of industrial pollution, the city did not apply any healthy environment policies to the five HI-TOADS examined in this research. Rather, through community empowerment policies at Nine Mile Run, the Hazelwood LTV Site, and the Heppenstall Site the city empowered NGOs to develop their own visions for the reuse of HI-TOADS and provided technical and financial assistance. At South Side Works and the Sears Site, the city's actions could be characterized as economic development.

While the City of Trenton applied a rational model considering location, market value, environmental conditions, and infrastructure when deciding which set of policies it would apply to which site, the City of Pittsburgh's decision-making seems to have been based more on politics and turf-battles than rationality. The city made little effort to apply policies systematically for HI-TOADS. For example, how can city officials justify to East Liberty residents that the city supported early reuse planning at other sites but not at the Sears Site?

Throughout my interviews with public officials and community representatives, I constantly heard the refrain "because Pittsburgh is in a weak market economy" to explain why development of HI-TOADS is such a challenge. Maxwell King (2006), president of the Heinz Endowment, recently reiterated the sentiment: "Pittsburgh has suffered in recent years from a weak-market economy" (1).

The Center for Economic Development (2002) at Carnegie Mellon University studied how Pittsburgh measured up in terms of competitiveness with other U.S. cities through over a dozen ranking reports. They found that Pittsburgh consistently ranked in the middle to upper-middle of most of the ranking reports. These finding demonstrate that the city has a strong real estate market relative to other Rust Belt cities and is able to attract modest levels of investment in the city, but that Pittsburgh is not a top-tier market. A 2005 Urban Land Institute report showed that Pittsburgh rated fair for investment and development of commercial and multifamily residential real estate (Miller 2005). This rating is far better than most of the cities included in the study results. It is also far better than how Trenton has performed.

Where the Trenton case study pointed out the need for wholesale public development of HI-TOADS because private investors were nowhere to be found, in self-declared weak-market Pittsburgh, the economics of development are much more favorable to HI-TOADS redevelopment. That is, for some HI-TOADS, land values are high enough that only a modicum of public subsidy can turn a negative financial assessment of a real estate project into a positive one. As a result, sites with little market value are not prioritized for public resources.

In discussing the development at South Side Works, Mark Knezevich of the URA said, "Without public funds [at the early stage of environmental remediation], it never would have happened. Otherwise, it would've stayed vacant. We're able to raise public funding when no one else can. When the public sector controls it, they can raise the kinds of funding to turn it into a developable site." By subsidizing a waterfront site in the midst of a booming neighborhood, the URA took on some risks, but primarily pushed the development costs lower. One community representative said, "I don't think South Side Works could have happened if South Side wasn't already considered a hot, happening place."

It was a similar story at Nine Mile Run. According to a URA official, "we thought it would work because of the strength of the surrounding neighborhoods." A former official commented, "there was a new market for new houses in Summerset." It seems that the URA would have never taken on South Side and Nine Mile Run projects if the sites did not have high market potential.

This appears especially true in contrast to the URA's thinking on Hazelwood. A URA official explained: "We felt the property had a negative value. We didn't go after it because we thought it was upside down, negative value. That the environmental clean-up costs exceeded value." When I asked a local community activist about success at South Side Works as compared with the lack of city efforts at other HI-TOADS, he responded "South Side, ha! It's a very strong market."

Pittsburgh's status as a weak market economy means that there are pockets of strong real estate values in the city and efforts over the last decade to address HI-TOADS have been focused largely in those areas. The only exception to this is the unusual city involvement at the Sears Site. I will explore this anomaly later in the chapter.

Pittsburgh has not updated its Comprehensive Plan since 1945. Despite that, officials in Pittsburgh have been involved in much planning since then (using the definition of planning introduced in chapter 2). Over the last decade, in particular, these efforts to address HI-TOADS have been comprehensive, future-oriented, and visionary. These efforts could be classified as exemplary government planning if they were not missing two key elements: a public process and codification.

The challenge faced by the Murphy administration was vast. One former official explained: "Pittsburgh has had 50 years of decline . . . we needed to invest in the future . . . [After coming into office] immediately we created the Pittsburgh Development Fund, shifted $7,000,000 into it, challenged corporations . . . [to contribute] $40,000,000, it grew to $60,000,000." Mr. Knezevich said: "When Mayor Murphy came into office in 1993, we acquired over 500 acres of brownfields sites" using that Development Fund. Of those 500 acres, 360 were HI-TOADS. A former city planning official felt that the Murphy administration undertook "a real attempt to get the community to reimagine itself." The effort involved physically recreating a large portion of the city's land from steel and iron production facilities to a new set of uses.

Future uses of HI-TOADS were determined largely by the proposals of developers, in response to URA-issued requests for proposals. But, generally, the city "developed an informal focus for each neighborhood" from a land-use perspective. A Downtown Plan and a Riverfront Plan were adopted during the last ten years delineating a certain range of new uses for those two zones. Both plans involved major public outreach processes and were adopted by the City Planning Commission. While the map of HI-TOADS in Pittsburgh shows the congregation of HI-TOADS along the river, the new plans were largely removed from the day-to-day deal-making that guided much of the URA's efforts (figure 6.1).

Part of what makes the Pittsburgh story of HI-TOADS reuse so fascinating is that city officials were focused closely on the highest and best use of their vast acreage of obsolete industrial lands. One of the chief aims of public land-use planning and zoning is to focus public resources to ensure that a mix of land uses can operate within a city and that lands with the greatest inherent locational or other advantages are reserved for value-maximizing uses (Babcock 1966). In a growing community, this can be accomplished through zoning and subdivision-control regulations whereby local officials can assess each property owner's proposed use. In a declining community, such tools have little utility. In Pittsburgh, the city government took an active stance to address a filtering effect.[5] Rather than allow these bottom-feeders to re-occupy what had been the locations of major employment, the city took ownership of the sites. A former city official explained: "Public sector has to own or control them at least initially. The only way you can hope to get the kinds of quality that will really lead to genuine increase in value is if you essentially protect it from the lowest common denominators . . . Need to protect it from the bottom-feeders." She went on to say that for the bottom-feeders, "it is likely that they would only go to the nearest term profitability . . . they're short-sighted." In contrast, the city is future-oriented and through the acquisition of these HI-TOADS, they report that they can better control the future use. A URA official feels that this investment of public resources means that the "quality of life return to the city is better" than had the bottom-feeders continued to occupy HI-TOADS. If the sites had been occupied continuously, then they would not have been HI-TOADS. While HI-TOADS pose a threat to their communities because of their abandonment, Pittsburgh officials felt that bottom-feeders represent a more dire threat to the future viability of the city.

Despite all of the good work done by the City of Pittsburgh, only the Downtown and Riverfront Plans were adopted formally. The rest of the city's efforts were ad hoc, and to some extent shielded from public scrutiny. While the city did work extensively with the public in many arenas, when it came to HI-TOADS reuse planning, the public was largely shut out. In describing the pro forma public meetings the URA hosted for Nine Mile Run, one former official said "two hundred people there, they all wanted to keep it a slag dump. They were concerned about change." Rather than view a public planning process as potentially positive, Murphy administration officials made comments like "people are in love with process, not decisions and actions, we were really focused on action" and "it's funny, Pittsburgh doesn't really have a comprehensive plan." Why no new master plan? One former city planning official responded, "we were working too hard, too fast." For all that visionary, long-term, comprehensive planning to address HI-TOADS, the city lacked a public process and way to formalize that vision.

An important piece of the Trenton case study was the uncomfortable mixing of history and progress in the city's planning efforts for HI-TOADS. Likewise, Pittsburgh was eager to forget the physical remnants of its industrial heritage. In the Murphy administration's rush to acquire HI-TOADS, little effort was paid to preserving the deteriorating historic structures. At the three HI-TOADS reuse projects, South Side Works, Nine Mile Run, and the Sears Site, the spirit of historic preservation was entirely absent.[6]

Prior to the South Side Works development, "everything came down" according to a URA official. The entire site was treated as a blank slate, something that city officials appeared to be attracted to. One community leader in a dense, mixed-use neighborhood resents the city's record of directing their efforts and resources at creating entirely brand new communities, while existing communities are made to suffer: "It's aggravating to me that they are starting fresh, because we are starved. Not as many-high profile ribbon cuttings [in these existing communities]. City officials . . . want to recreate a neighborhood or business district . . . create another fiefdom." The result of "starting fresh" is palpable during tours of South Side Works and Nine Mile Run. In both these cases and in other examples of large brownfields in the city, the URA and city administration showed their predilection for developing large sites anew, without attending to historic preservation.[7] Due to the deteriorating condition of many Pittsburgh HI-TOADS and the extraneous costs often involved in historic preservation, the URA simply did not focus on historic preservation. More critical was the idea of breaking from the perhaps ugly past of some of these sites.

While historic preservation was absent from the successful city-led HI-TOADS reuse projects, planning underway for the Hazelwood LTV Site includes a small historic preservation element and efforts at the Heppenstall site in Lawrenceville include the adaptive reuse of some structures. A developer involved in the Heppenstall project explained why his organization was committed to historic preservation: "You have the emotional tie, where their father and grandfather worked, and then you have the people who are thrilled. We will be keeping some of the buildings, reusing the buildings."

Most unusual about the city's resistance to historic preservation at HI-TOADS is the vitriolic feelings expressed when Pittsburghers discuss the travesties that occurred through urban renewal more than fifty years ago. One community activist said of urban renewal, "these things all destroyed the neighborhood. As a result of urban renewal, public housing concentration, pedestrian malls, it destroyed that neighborhood." Yet today, the "historic" built environment is the subject of the same contempt as was the built environment of the 1950s. The East Liberty neighborhood was regarded as "obsolete" and labeled "slums" by city officials in the 1950s (Fitzpatrick 2000a). When city officials encountered the Sears Site, the fifty-year-old structure's historic significance was not even given a thought.[8] The city's rush to assist Home Depot in demolishing the site and subsidizing the construction of a new building shows an equally callous disregard for the potential historic value of buildings that URA planners of a half-century ago demonstrated.

The central questions for this study involved city governments as the central unit of analysis. What is it that city governments do, either directly or in support of other nongovernment organizations (NGOs)? What appears clear from the Pittsburgh case study is that for each of the HI-TOADS, it is the NGOs that catalyzed the reuse of HI-TOADS, not political or administrative leadership. Yet, the city government hardly seems to recognize this fact and its formal and informal planning and public policy relative to HI-TOADS effectively leaves NGOs out of the debate.

It was the South Side Neighborhood Development Corporation that first worked directly with LTV to craft a community-driven reuse plan at South Side Works. A neighborhood activist said, "we knew something was going to happen on the site . . . we got a grant from LTV to do the plan." This community planning exercise then led to a more formal public process, the South Side Planning Forum, which adopts a neighborhood plan every two years. These efforts culminated in the URA acquiring the site and directing a full-scale reuse of the HI-TOAD site.

It was a similar story at Nine Mile Run. A local activist explained: "I was involved from a neighborhood's point of view in a process to conceive the idea of using the slag dump at Nine Mile Run as residential development . . . By the time the first residents moved in, there were all kinds of elected officials taking credit for it, just so you know it really came from a community-based effort."

A nonprofit developer at the Heppenstall Site said, "we're an organization that precedes the private sector, we initiate, catalyze." With the foundations and their new organization, ALMONO, at Hazelwood, and the East Liberty Quarter Chamber of Commerce at the Sears Site, NGOs have been at the front and center of all HI-TOADS reuse efforts in Pittsburgh in the last decade.

In the recent past, the Department of City Planning was a proactive force for galvanizing the efforts of NGOs in Pittsburgh. Today, it is the NGOs that galvanize the city to act. The director of one such NGO summed it up well: "We used to have [Department of City Planning] neighborhood planners assigned to neighborhoods, it became the job of the CDCs to do that work . . . The vision, the tenacity, is all coming from the CDCs, the city is clearly a good partner, they have to go hand-in-hand." Ultimately, the NGO's catalytic involvement is not a neutral, rote involvement, but rather an impassioned, visionary, and tenacious involvement, as described above. The efforts of ALMONO at the Hazelwood LTV site are viewed by many in the city as a commitment by the foundations to quality development to improve the city. "[Foundations] said 'this is our last big tract of land in the city, let's do this right,'" according to a local activist. A former city official who often clashed with NGOs conceded their importance: "they have been advocates for the best, higher value use and, in fact, such a high value it was often unrealistic . . . In general, a force for good."

With all their energy and enthusiasm, the NGOs end up competing with the URA and the mayor for the glory of redeveloping HI-TOADS. A nonprofit developer commented to me: "We're on the outs with the mayor, he saw us as competition." In my observations, it is this very competition that has generated so many HI-TOADS redevelopment efforts. This wide array of NGO players, foundations, community-based organizations, CDCs, nonprofit developers, combined with a strong mayor and a powerful URA has meant that a lot of entities are trying to make their mark through HI-TOADS redevelopment.

The choices of the city to invest in high-market-value HI-TOADS means that the nongovernmental sector is left to tackle the proverbial "high-hanging fruit," the more difficult sites. But despite their competitiveness, much of the HI-TOADS reuse successes have been executed through partnerships. The model was as follows:

1. An NGO would instigate a planning effort to address a HI-TOAD site;
2. Because of the city's long-term vision for re-defining the city's land uses, funding and technical assistance were in place to act on the recommendations of the NGO's plan;

3. The URA would issue requests for proposals to seek a development partner while utilizing external county, state, and federal funding as well as the city's own funds to cover environmental investigation and remediation that it would manage;

4. With some public input, the URA and its development partner would finalize a reuse plan and execute it.

This model, however effective in turning three of the city's HI-TOADS into productive uses, lacked a formal mechanism for tying the reuse of each HI-TOAD site into a large, more holistic plan for the city, and for encouraging the participation of the public and NGOs in a proactive plan. This ad hoc, informal planning approach can systematically exclude the less powerful NGOs and marginalize concerned citizens from important decisions about which sites should be prioritized for redevelopment, how should they be reused, and who should lead the reuse.

First Whales, Now Brownfields

NEW BEDFORD, MASSACHUSETTS

L ocated in southeastern Massachusetts, less than 60 miles from Boston, New Bedford was a great industrial center of the nineteenth century and has struggled ever since. With high unemployment, high crime, and high poverty levels, New Bedford would seem to be just like so many other Rust Belt cities that lost their job base and are seeking the next big thing to save them. But that is not New Bedford's story. In fact, New Bedford has a thriving job market. Immigrants from throughout the world, especially the Portuguese-speaking world, flock to the "Whaling City" and find low-skill work with industrial firms.

After studying Trenton and Pittsburgh, with both cities hostile to bottom-feeder uses (that is, low market value uses like junkyards and recycling storage areas, which create few local jobs), I was struck by how widespread bottom-feeders are in New Bedford. From a HI-TOADS perspective, bottom-feeders occupy otherwise vacant buildings, protect them from fire and criminal activity, and hence ensure that the negative externalities generated by HI-TOADS do not spill into neighborhoods. New Bedford's embracing (passive or otherwise) of bottom-feeders is central to this case study and to understanding the larger questions of this study.

Like both Trenton and Pittsburgh, New Bedford has received recognition for its efforts to address brownfields. In 2001, the city was designated a Brownfields Showcase Community by the U.S. EPA. But the city's leadership and activism in this arena was devoted mostly to high-market-value brownfields that did not have significant negative pull on neighborhood property values, the proverbial "low-hanging fruit."

Key Questions for New Bedford

In this chapter, I explore the ways that bottom-feeders have been viewed in New Bedford by local officials as a natural market response to HI-TOADS. Genuine demand exists for large abandoned sites for low-end uses that would not be tolerated by city officials in Trenton and Pittsburgh. But in New Bedford, the bottom-feeders are tolerated. Why?

A second question for the New Bedford case study is to explore further the role of the federal government in supporting HI-TOADS reuse. Like Trenton, New Bedford had a federal employee located in City Hall to support brownfields redevelopment. My telephone interviews with two local officials in New Bedford revealed that federal support was seen by both as central to the city's efforts at HI-TOADS. This was in contrast to the perceptions of Trenton officials, who looked at federal support as useful but not the most important independent variable in determining their successes.

In 1996, Congress designated 34 acres of New Bedford's downtown as a National Historic Park (National Park Service 2006). This extraordinary commitment to historic preservation has had wide-reaching impacts on the city. Given the National Historic Park's precedence and the city's growing interest in promoting tourism, what role has historic preservation played in the city's efforts to address HI-TOADS?

Herman Melville made New Bedford famous in the classic tale, *Moby-Dick* (1851). Seventy-five years after the book's publication, the last whaling boat left New Bedford harbor (Wolfbein 1944). This one-industry town suffered greatly when oil reserves were discovered in Pennsylvania and the market for whale blubber vanished. But the city's attachment to the sea has persisted until today. The traditions of whaling and maritime trades continue to be a part of the city's heritage, most clearly demonstrated in its working waterfront.[1] Despite the difficulties caused by widespread unemployment after the crash of the whaling industry, plans were in the works by the middle of the nineteenth century to invent a replacement business for the city. Many individuals and families grew wealthy during the height of whaling and invested those savings in cotton textile mills. The first cotton mill in New Bedford was erected in 1846. By 1932, there were 42, and "the city had become the center of the manufacture of fine cotton goods in the United States" (Wolfbein 1944, 10).

In the same way that the bottom fell out of whaling, so too did the cotton market crash. By the end of the Great Depression, the lack of industrial diversity was quite apparent. "It is probable that at least 60,000 persons, or one-half of the entire population were directly and immediately affected by the collapse in cotton textiles" (Wolfbein 1944, 107). The decimated New Bedford community sought to fight against this widespread unemployment and consequential poverty, but the regrowth of an employment base in New Bedford has been a slow process.

The widespread availability of federal urban-renewal monies was welcomed by local New Bedford leaders in the 1950s and 1960s. In a series of demolitions and redevelopments, city officials sought to recreate portions of the city's neighborhoods. An unusual facet of this story is that a small band of historic preservationists were successful in utilizing federal urban-renewal monies for historic preservation. This group asked, "How can we save our city's heritage from death by neglect and the tyranny of the bulldozer?" (McCabe and Thomas 1995, 10). The group, the Waterfront Historic Area League (WHALE), believed that "the rehabilitation of the waterfront could raise the spirits of the whole city" (McCabe and Thomas 1995, 25).

The efforts of WHALE eventually led to the 1996 designation of the New Bedford Whaling National Historic Park covering 34 acres at the heart of the city's historic center. The urban renewal, historic preservation, and federal park designation all combine to make downtown New Bedford a unique place today. The city was profiled in a recent Escapes section of the *New York Times*. "Tough times and a rough reputation is how the city is generally perceived regionally . . . Truth is, though, New Bedford has plenty of history, architecture, and small museums to fill a weekend" (Schneider 2006, 3).

Part of that legacy that tourists may not appreciate is the widespread contamination of the New Bedford Harbor and Acushnet River with polychlorinated biphenyls (PCBs) in the 1970s. The dumping of PCBs into the city's waterways by manufacturers of electronic components has put the city in the unpleasant spotlight as hosting the most polluted harbor in the nation (Voyer et al. 2000). Efforts have been ongoing for the last decade to remedy the situation and repair the ecological health of the city. In many ways, the story of HI-TOADS redevelopment is part of a process in New Bedford to recover from its notorious environmental degradation.

HI-TOADS in New Bedford

There are five HI-TOADS in New Bedford: Aerovox, the Elco Dress Factory, Fairhaven Mills, Morse Cutting Tools, and Pierce Mills (see table 7.1 and figure 7.1).

In this section of the chapter, I will describe three of the New Bedford HI-TOADS in great depth (Elco Dress Factory, Morse Cutting Tools, and Pierce Mills) in order to highlight the range and types of sites in the city. Next, I will provide background on reuse and redevelopment efforts at each, and present the results of statistical analysis of census data for each site. I will then synthesize the data analysis for all five HI-TOADS and present a summary of what I found with respect to HI-TOADS redevelopment, socio-economic changes, and property values in New Bedford.

Elco Dress Factory

Constructed in 1909, the Elco Dress Factory was a booming enterprise for almost a century in the city's historic North End neighborhood. Sandwiched between industrial and residential uses, the 50,200-square-foot, four-story structure is an imposing presence on its neighborhood (see figures 7.2 and 7.3). Abandoned for over a decade, the property came into the hands of the City of New Bedford through nonpayment of back taxes.

Like so many other buildings of its type, a modern industrial reuse of the site would be challenging. The city completed a Phase I Environmental Site Assessment in 2000 and found no other "recognized environmental conditions" at the site, except for the likely presence of lead-based paint and asbestos (City of New Bedford 2006).

In the six years since the completion of the environmental assessment, no physical movement has occurred on the property. But just this year, the state has provided the city with a $200,000 grant to demolish the building. In the city's press release announcing the grant, City Councilor Paul Koczera was quoted as saying, "I am extremely happy that the wishes and concerns of the residents of Taber Mill and the neighborhood are finally being addressed."

Taber Mill is a three-story adaptively reused mill building recently converted into affordable elderly housing. Just a block away from the Elco Dress

Table 7.1

Summary Information for HI-TOADS in New Bedford

Site name	Size	Location	Historical use	Proposed use	Status of redevelopment efforts (as of 2006)
Aerovox Site	10.5 acres	740 Belleville Avenue	Electronics manufacturing	Industrial	Site remediation underway.
Elco Dress Factory	1.5 acres	330 Collette Street	Dress manufacturing	No plans in place	Preliminary site investigation undertaken.
Fairhaven Mills	3.6 acres	85 Coggeshall Street	Mill operations	No plans in place	Partial demolition occurred in early 2000s.
Morse Cutting Tools	3.4 acres	163 Pleasant Street	Cutting tools manufacturing	No plans in place	Demolition occurred in 1997, site remediation on going.
Pierce Mills	3.5 acres	226–378 Belleville Avenue	Mill operations	Neighbor-hood park	Demolition of derelict structures occurred in mid-1990s, park opened in 2003.

SOURCE: New Bedford Chamber of Commerce (1997); personal interviews with local officials.

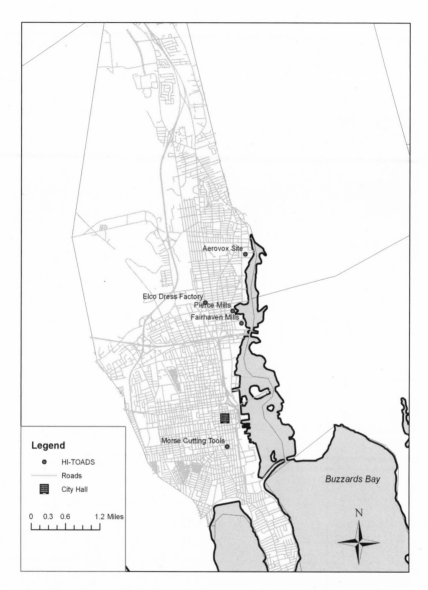

FIGURE 7.1. HI-TOADS in New Bedford, Massachusetts.

FIGURE 7.2. Elco Dress Factory: Looking east along Collette Street.

site, Taber Mill residents have been vocal about abandoned buildings in their neighborhood. Last year, Taber Mill residents were credited with convincing the city to demolish another nearby abandoned building, Payne Cutlery, on Church Street. If the demolition of the Elco Factory goes forward as planned, once again the Taber Mill residents could be seen as the instigators.

Among the three broad categories of policy responses, the city's actions at Elco Dress fit into the community empowerment set of policies. The decision to advance environmental remediation at the site and to request demolition money from the state were not part of a larger strategy to attract new residents or generate new jobs (as are typical of economic development), but rather occurred primarily at the urging of local residents fearful of the potential for a fire at the abandoned building. The city's efforts to attack the perceived threat to the health and well-being of Taber Mill residents is also exemplary of healthy environment policies. Because the Taber Mill residents were concerned about the hazards of an abandoned structure—a public health issue—their demands were palatable to city officials.

The Taber Mill residents' ability to effectively organize, develop a strategy, and deliver that strategy to the city's political leaders exemplifies community empowerment. Missing from the city's efforts here is

FIGURE 7.3. Elco Dress Factory: The site overwhelms the neighborhood. Looking south between two multi-family homes along Davis Street.

a codification of local visions for how the Elco site can be reused. At this point, the city is only willing to support the demolition project and has not committed to any form of public planning process to explore reuse possibilities for the site.

While local officials agreed that Elco fits the definition of a HI-TOAD site, the census tract did not fare much differently between 1980 and 2000 than the whole of New Bedford (see table 7.2). The mean housing value in the tract was only $4,400 per unit in 2000 (compared with a mean value citywide of $31,400 per unit). Average household income and owner-occupied housing levels are roughly the same as the city as a whole. While the Elco site may have had a dragging effect on neighborhood property values, the census data show that the area grew and prospered from 1980 to 2000 in terms of housing appreciation, growth in rents, and growth in owner-occupancy. The neighborhood's population did fall 4 percent during this period, the same rate that the city's overall population dropped.

Table 7.2

Property Value and Demographic Data for HI-TOADS and the City of New Bedford, 1980 to 2000

	1980	1990	2000	Percent change		
				1980–1990	1990–2000	1980–2000
City of New Bedford						
Mean housing value ($1,000s)	8.8	30.8	31.4	250	2	257
Median gross rent ($)	187	455	404	143	-11	116
Percent rental units	58	56	56	-3	0	-3
Average household income ($1,000s)	11.60	21.90	29.40	89	34	153
Population	98,127.00	99,922.00	93,768.00	2	-6	-4
Percent African American	3	4	7	42	80	156
Percent Hispanic	5	6	10	20	79	115
Aerovox Site						
Mean housing value ($1,000s)	13.60	39.00	36.80	187	-6	171
Median gross rent ($)	224	443	508	98	15	127
Percent rental units	46	47	49	2	4	6
Average household income ($1,000s)	15.50	33.80	43.90	118	30	183
Population	3518	3451	3213	-2	-7	-9
Percent African American	0	0	1	—	126	—
Percent Hispanic	2	1	1	-77	2	-76

Table 7.2 (*continued*)

	1980	1990	2000	Percent change		
				1980–1990	1990–2000	1980–2000
Elco Dress Factory						
Mean housing value ($1,000s)	1.6	4.7	4.4	194	–6	175
Median gross rent ($)	188	368	445	96	21	137
Percent rental units	72	72	68	1	–6	–5
Average household income ($1,000s)	11.10	24.20	30.00	118	24	170
Population	3677	3981	3529	8	–11	–4
Percent African American	0	1	5	62	557	961
Percent Hispanic	2	4	10	151	146	516
Fairhaven Mills						
Mean housing value ($1,000s)	0.6	6.5	2.8	983	–57	367
Median gross rent ($)	183	396	457	116	15	150
Percent rental units	73	65	72	–11	10	–2
Average household income ($1,000s)	12.7	21.9	20.9	72	–5	65
Population	2000	1926	1669	–4	–13	–17
Percent African American	2	3	11	56	262	463
Percent Hispanic	8	10	27	17	186	233

Morse Cutting Tools

Mean housing value ($1,000s)	4.8	25.4	25.3	429	0	427
Median gross rent ($)	188	420	463	123	10	146
Percent rental units	54	52	55	−4	5	1
Average household income ($1,000s)	14.50	30.00	43.20	107	44	198
Population	2998	3024	2842	1	−6	−5
Percent African American	4	4	8	−16	129	91
Percent Hispanic	8	2	5	−79	232	−29

Pierce Mills

Mean housing value ($1,000s)	0.4	3.3	2.5	725	−24	525
Median gross rent ($)	173	418	456	142	9	164
Percent rental units	80	74	70	−8	−6	−13
Average household income ($1,000s)	10.2	19.8	24.3	94	23	138
Population	2534	2461	2256	−3	−8	−11
Percent African American	0	2	7	—	314	—
Percent Hispanic	7	7	23	15	202	246

Morse Cutting Tools

Situated in the heart of New Bedford's primarily residential South Central neighborhood, the Morse Cutting Tools plant contributed to the livelihood of hundred of residents and was seen as a center of the community. "I don't think there was a family in the neighborhood without someone in their family who worked there," observed a community leader. When the plant closed, the owners walked away and left a tremendous environmental mess behind, highlighted by the spillage of vast quantities of lubricating oil into the soil. In describing the structure prior to its demolition, one former official said "it would weep oil onto city streets." That oil would then be seen "draining down the sidewalk."

For decades, community groups demanded that the city help them address this dreadful visual problem and public health threat, but to little avail. Buddy Andrade and his organization, Old Bedford Village, and John Simmons with Hands Across the River were both enlisted to help the neighborhood to understand the environmental challenges they faced and demand action from the city.

Despite the city's efforts to control the process of addressing the Morse site, one famous public neighborhood meeting ended with residents chanting "tear down the building." According to a city official who was present, "it became apparent, at that point, that we had to tear down the building."

In an on-site ceremony in 1996, Lt. Governor Paul Cellucci and Jane Gumble of the state Division of Housing and Community Development presented a mock check for $700,000 to cover the demolition of the Morse structures. The money came from a now-defunct State of Massachusetts fund that provided direct grants to local governments to pay for demolition of abandoned structures. As new priorities emerged within state government in the late 1990s and early 2000s, support waned and the fund was eliminated.

As he took a swing at the building with a sledgehammer, Mr. Cellucci told the assembled crowd that this demolition work would "remove a severe public safety and environmental threat to this community" (as quoted in Corey, Stewardson, and Thomas 1996). Despite the long line of politicians in attendance at the ceremony, they all "laid credit for the award squarely with the people of the South Central neighborhood." One even remarked, "I believe the residents are the heroes" (Corey, Stewardson, and Thomas 1996). The remainder of the structures were demolished completely in 1997.

A collaborative effort between the city, activist neighborhood groups, and the state led the state attorney general to identify the media-giant Viacom as a potentially responsible party for the site's contamination. While the derelict structures are gone, ongoing environmental remediation will continue on half of the site for a few more years. Viacom's successor, CBS, is now paying for that remediation work. The other half of the site has been grassed over and is maintained by the city park's department (see figures 7.4 and 7.5). Despite all of this progress, no plans have been finalized for the site's reuse; the city views the grassed-over parcel as an interim use.

The city's efforts at Morse can be classified as both community empowerment and healthy environment policies. The city's interventions focused primarily on removing the imminent threat to public health and the environment. Secondly, though, the city's engagement with neighborhood groups and the public went far to fulfill the needs of the community and to better empower CBOs and local residents to work toward a positive change in their neighborhood. The lack of a clear reuse plan for the site in some ways is attributed to the city's failure to engage *fully* in a community

FIGURE 7.4. Morse Cutting Tools: Structures at the site have been demolished. The westerly parcel is grassed over and is used for passive recreation. A view along South 6th Street looking south.

FIGURE 7.5. Morse Cutting Tools: The easterly parcel is still undergoing remediation. It is improved by a temporary structure and fenced off. The view is from Pleasant Street looking east.

empowerment framework. Such an effort would have better positioned the community groups in the neighborhood within the decision-making process for the demolition, remediation, and reuse of the site. As it happened, the city's efforts did not facilitate or empower a community-driven reuse process, at least not yet.

In examining census data, I am using 1997 as the date when the site was no longer a HI-TOAD (according to my findings from the interviews). Therefore, the key range of dates is from 1990 to 2000. During that time period, the Morse census tract changed in housing value growth at roughly the same rate as the city as a whole. But the tract experienced a 10 percent growth in rents while the city experienced an 11 percent drop in rents during the period. And despite the drop in population (5 percent), average household income grew from $30,000 to $43,200, 47 percent higher than citywide income levels. While certainly not the only factor, the elimination of the Morse Cutting Tools HI-TOAD site may have made a positive contribution to enhancing the wider South Central neighborhood.

Pierce Mills

Closed and abandoned in the mid-1980s, the hulking Pierce Mills complex was a visual barrier between the Bullard Street Neighborhood and the Acushnet River. Typical of HI-TOADS, the site was an eyesore and attracted repeated arson (Corey 1996). To many, it was seen as source of blight on the neighborhood.

Under the leadership of the Bullard Street Neighborhood Association and Hands Across the River, local politicians were informed that residents wanted the HI-TOAD site to be demolished and replaced with a park. Normally, such a request would have been considered a pipedream, but a financial settlement with polluters under the EPA Superfund program had just created the Harbor Trustee Council with the charter of dispensing monies to "to improve sections of the Acushnet River spoiled by the pollution" (Nicodemus 2003, A9). The dumping of PCBs into the city's waterways had made New Bedford's harbor the subject of an intense legal battle. The ultimate settlement and establishment of the Harbor Trustee Council and its grant program was viewed as a way to remedy past mistakes (Voyer et al. 2000).

As opposed to the top-down, mayor-driven brownfields projects in New Bedford that garnered so much attention in the 1990s, the Pierce Mill effort was bottom-up. According to a city official, "it was a community decision to turn it into a park." And because the Harbor Trustee Council had funding available for such a purpose, the local voice was heard and the city followed through (see figure 7.6). At the ribbon-cutting ceremony when the park was opened, Mayor Fred Kalisz was credited with "shepherding the project though to completion" (Nicodemus 2005, A1).

As at Morse Cutting Tools and the Elco Dress Factory, the city's efforts at Pierce Mill also fit into community empowerment and healthy environment policy sets. An initiative by local residents to turn an eyesore into a community park was both facilitated by and "shepherded" by city leaders. The provision of open space and fresh air to an entire neighborhood had a multitude of public health and environmental benefits for area residents. City policies were oriented around these goals and were ultimately successful.

Because the demolition of Pierce Mill occurred in 1996, I will be examining census data for the period from 1990 to 2000 to note any changes in the census tract that hosts Pierce Mill. Table 7.2 shows the astronomical growth

FIGURE 7.6. Pierce Mills: The recently opened park is improved by soccer fields and a playground. A wrought iron fence and antique streetlamps line Belleville Avenue.

in housing values from 1980 to 1990 for the tract (725 percent) and then the drop in values by 24 percent from 1990 to 2000. But, rents did rise in the tract by 9 percent from $418 to $456, above the average for the city. While the site was no longer a HI-TOAD by 1996, much uncertainty surrounded the reuse of the site. In 2003, an article in the New Bedford newspaper, the *Standard-Times,* led with a headline: "City assures that Pierce Mill will be a park" (Nicodemus 2003). The concern among the neighbors of the site was very real for some time after the demolition. Rumors were circulating that the site might be returned to industrial use, which may have played a role in keeping down property values for the surrounding residential neighborhood.

How Are HI-TOADS Addressed in New Bedford?

Like Pittsburgh and Trenton, over the last decade city officials in New Bedford have seen their role as developers of brownfields. But in New Bedford, the city's interest in developing and reusing HI-TOADS, specifically, has been a low priority. The city's brownfields policy was oriented toward leveraging outside resources and job creation. Fred Kalisz was first elected into the mayor's office in 1998 and remained there for four two-year terms. Like

Mayor Murphy in Pittsburgh, Mayor Kalisz focused much of his administration on brownfields redevelopment. Also like Mayor Murphy, Mayor Kalisz developed a development fund to support his efforts at remaking New Bedford.

Through a three-pronged approach, Mayor Kalisz focused on neighborhood revitalization, job creation, and the development of New Bedford as a multi-modal transportation hub. One former official in the Kalisz administration explained: "The mayor's job is multifaceted . . . you try to bring vision for the future and clean-up from the past."

This vision and some key successes on the ground gained Mayor Kalisz a reputation in the brownfields community as a leader. While mayor of New Bedford, he served as president of the Massachusetts Chapter of the National Brownfields Association.

The city's approach began with an inventory of all brownfields, an effort funded by the New Bedford Chamber of Commerce. At first, all potential brownfields were considered, but after some consideration, the analysis honed in only on city-owned brownfields, 40 in total. Then, working with a Brownfields Task Force comprised of city, state, federal, and community leaders, the sites were slotted into one of three reuse categories: commercial/industrial, transportation-related, and residential/recreational. City officials explained that commercial/industrial uses were planned for sites wholly within existing commercial/industrial districts.[2] In instances where a site's prior industrial/commercial use was not compatible with surrounding land uses, the site was slotted into a residential/recreational category. For all sites that offered a potential to enhance New Bedford's standing as a major intermodal transportation center for the region, such a transportation reuse was planned.

A former official described their effort using the language of neighborhood life-cycle theory: "Urban environments have not been maintained over the years. We were aggressive strategic surgeons, otherwise, patient dies." When it came to applying this aggressiveness to HI-TOADS to the sites that presented the greatest threat to neighborhoods, the city's efforts fell flat. The success at Morse Cutting Tools was largely due to a combination of citizen pressure and a state demolition grant. The "slotting" process employed by the Brownfields Task Force was ineffective there because consensus could not be reached on how the site should be reused. The Pierce Mills effort likewise was instigated by citizens groups and paid for by the

Harbor Trustee Council.[3] The other two HI-TOADS in New Bedford not profiled in this chapter, Fairhaven Mills and the Aerovox Site, have been the subject of attention of the city but despite those efforts remain HI-TOADS. At all of the HI-TOADS in New Bedford, the city acquired the sites under tax foreclosure and has been attempting to advance environmental investigation, demolition, and redevelopment in partnership with a private developer. A request for proposals was issued in 2006 for the Aerovox Site, but reuse and redevelopment could still be years away.

While I examined counties in the Trenton and Pittsburgh case studies, Massachusetts has no county form of government; therefore I will focus this section on what it is that the Commonwealth of Massachusetts and the federal government do about HI-TOADS.[4] The state played a key financial role in paying for the demolition of the Morse Cutting Tools buildings and will finance the demolition of the Elco Dress Factory. The Division of Housing and Community Development's Demolition of Abandoned Buildings Program covered the cost of demolishing the two structures at Morse Cutting Tools. An official with the Massachusetts Department of Environmental Protection summed up how important the state is to local governments' efforts to address HI-TOADS: "They need public money to make it work."

But the state also has provided technical support for the city's efforts to remediate the Aerovox and Pierce Mills sites. Massachusetts' Department of Environmental Protection has a statewide brownfields program, with field officers in each region. The result has been a close working relationship among regulators and city officials where creative solutions are found for challenges. One state official explained, "we're at 'a,' have to get to 'b' . . . we make the regulations work for you."

The federal role in New Bedford has been fairly substantial. The contamination of New Bedford Harbor with PCBs essentially put the city on the radar screens of the EPA and the National Oceanic and Atmospheric Administration (NOAA). The federal action in prosecuting responsible polluters led to the creation of the Harbor Trustee Council, which contributed to the reuse of Pierce Mills through its funding commitment. The EPA's designation of the New Bedford Harbor as a Superfund site led NOAA into planning and remediation activities and eventually into its designation of

New Bedford as one of three Portfields pilot cities in the United States. NOAA's Portfields initiative focuses on promoting the redevelopment of brownfields in port and harbor areas, "with emphasis on development of environmentally sound port facilities" (National Ocean and Atmospheric Administration 2006).

In 2001, EPA designated New Bedford as a Showcase Community and further federal resources flowed to the city, including the relocation of a federal employee to City Hall for full-time work on brownfields. While EPA employees typically were assigned for these positions, in New Bedford, NOAA provided one of their own employees, Robert Neely. For three years, Mr. Neely worked closely with Scott Alfonse and the city solicitor, Matthew Thomas, to expedite the development of brownfields sites in New Bedford. According to a former officials: "[Mr. Neely] worked with the mayor and Scott [Alfonse] to make sure they were plugged into federal resources." He was described by a former city official as "invaluable" to the city's efforts. Around City Hall, they jokingly called Mr. Neely "the government agent."

Despite all of his focus on brownfields policies, Mr. Neely's efforts were not aimed directly toward any of the city's HI-TOADS. Thus, his contribution was largely in helping city officials develop strategies for the reuse of sites and getting them connected to sources of federal funding. During his tenure with the city, Mr. Neely worked on brownfields identified by the city as priorities, none of which were HI-TOADS.

The not-for-profit nongovernmental sector in New Bedford is active for a city of its size. This activism may be due in part to how the EPA managed plans for a clean-up of the New Bedford Harbor Superfund site in the late 1990s. With only cursory public input, the EPA decided to dig up PCB-laden soil from underwater, store it along the Acushnet River, and then burn it. Intense public outcry (led by the NGO Hands Across the River) followed the announcement of these plans and a record of strong public participation in New Bedford has followed that debacle (Schattle 1998).

In this study of HI-TOADS, I found that vital NGOs were linked to four of the five sites. The Old Bedford Village, Hands Across the River, and Bullard Street Neighborhood Association were all active in instigating the Pierce Mill reuse into Riverside Park. Old Bedford Village and Hands Across the River also were involved, with the South Central Neighborhood Association and the Minority Action Coalition, at the Morse Cutting Tools site. At the Elco Dress Factory, the organized residents of Tabor Mill have played an important role in influencing city efforts there. And the historic preservation controversy at Fairhaven Mills brought the preservation group WHALE into the public debate. City officials indicated that once reuse planning for the Aerovox Site begins in late 2007, the local neighborhood group will be engaged formally. Up until now, there has been no formal involvement of an NGO at Aerovox.

The private, for-profit sector was conspicuously absent from HI-TOADS reuse in New Bedford. Over the last several years, private developers advanced a reuse plan for Fairhaven Mills, but ultimately, that project did not advance. While new developments at other brownfields sites have been funded through private investment, the sites included in this study were passed over. As evidenced by the city's Brownfields Task Force, the city's private, for-profit sector is interested and engaged in redevelopment. Due to tax delinquency and other factors, New Bedford's HI-TOADS are owned by the city, a factor that may drive developers disinterest in acquiring and redeveloping them.

Success in this study can be measured by whether a HI-TOAD site is seen by local officials and community leaders as having a negative pull on surrounding property values. When it no longer does, then it is no longer a HI-TOAD site and success is achieved. A city official expressed his concurrence with that logic: "You have a reused site, that might be the biggest metric you need." In this study of New Bedford over the last decade, two of the five HI-TOADS were converted successfully (Morse Cutting Tools and Pierce Mills).

While the creation of Riverside Park at the former site of Pierce Mills is an indisputable improvement for the neighborhood and for the city, the end-state land use at Morse Cutting Tools is unknown. With such uncertainty in terms of future use and because of the acrimonious debate surrounding the demolition of the buildings, the legacy of the Morse Cutting HI-TOAD site will be long-lasting.

On a positive note, recent progress at the three other HI-TOADS, Aerovox, Fairhaven Mills, and the Elco Dress Factory suggests that the city and neighborhood groups are on a course to redeveloping each site. For each, city officials have identified resources and a path to begin the process of reusing them and the wide economic climate in the city has improved such that demand for the sites is growing. It is likely, therefore, that those sites will no longer be HI-TOADS a few years from now.

Key Themes in New Bedford HI-TOADS Redevelopment

In this section, I move beyond describing the conditions in New Bedford and the efforts to address HI-TOADS to distilling the key themes from the case study. As in the Trenton and Pittsburgh case studies, I present below five themes that arose from the analysis, corresponding loosely with five questions that have driven much of the case-study work.

Despite the city's stated brownfields policy to generate new jobs, there was little in the city's efforts toward HI-TOADS that fit into the economic development category. While many of the city's efforts at the smaller, less-noxious brownfields sites could be described as exemplary of economic development, the city's strategies toward HI-TOADS bore a much closer resemblance to community empowerment and healthy environment policies.

The Pierce Mills conversion to a park showcased the city's commitment to both community empowerment and healthy environment policies. While not featured in the chapter, the Aerovox Site was closed by the U.S. Environmental Protection Agency because of health risks to workers. The city also drew on health environment policies to relocate that business and manage the remediation and redevelopment of the site, which is ongoing.

The failure of the Fairhaven Mills economic development-oriented reuse and the successes at Morse Cutting Tools, Pierce Mills, and Elco Dress Factory suggest that community empowerment worked better in New Bedford. The failure of the Kalisz administration to operate transparently and to involve high levels of public participation in decision-making may have contributed to an environment of mistrust, where the only successful HI-TOADS reuse are those driven by a community-based organization.

HI-TOADS were not part of the city's planning efforts. But, formal efforts at government planning were not common in the first place in New Bedford. Without an update to its master plan since the 1960s, the city's formal planning has been oriented toward the harbor: a 2002 Harbor Master Plan and a 1999 Harbor Open Space Plan. Neither plan includes direct references to any of the HI-TOADS included in this study.

City officials argued to me that despite the lack of a formal planning process, the city had undergone comprehensive, proactive, visioning that fits into a less-formal definition of planning. At the centerpiece of this effort was the way that the city, in close coordination with a brownfields task force, "slotted" properties into one of three categories: commercial/industrial, transportation-oriented, and residential/recreational. Despite the failure of this effort to advance HI-TOADS to redevelopment, it is an example of planning. However, it fails the basic rule of a governmental planning process, which is a public participation component. This reuse planning was done without public comment and was never codified and disseminated to the public.

But interviewees felt that the lack of a more formal planning process with full citizen participation was a real missing link in the city's efforts to address HI-TOADS. "The city doesn't have a master plan, you use that as a foundation for all decision-making," remarked one city employee. Reuse planning occurred on an ad hoc basis, more reactionary than anything else. This lack of a master plan, this ability to act on an ad hoc basis, just may be the preference for city leaders.[5]

In interviews with current and past city officials, I gained insights into their motivations that contrasted greatly with what I learned in Trenton and Pittsburgh. In New Bedford, the city took up activist brownfields policies because of two major factors, the genuine economic development opportunity presented by brownfields, and city officials' sense that the emerging national brownfields movement could be used to their political and fiscal advantage.

New Bedford is almost completely built-out. For the city to experience an expansion of its job base or property tax revenues to meet increasing costs, the only hope they have is for development at previously developed sites (some of which are HI-TOADS, many of which are brownfields). A City of New Bedford official described brownfields and HI-TOADS as the "single most important redevelopment issues in the city . . . New Bedford is 95 percent built out. The only way to grow and change economic forces in the community is brownfields."

This development challenge becomes a true public problem because such a large percentage of those sites are city-owned. The city had to devise a strategy not simply to position the city for economic growth, but to lead such an effort as an active agent. As the owner of what came to be estimated at forty brownfields, the city may have been the largest landholder of brownfields in New Bedford. The logistic and administrative challenge of disposing of so much property (much of it with negative value) was a central driver for the city's efforts.

But many cities in the deindustrialized Northeast and Midwest faced a similar predicament. The second factor that made New Bedford unique was the ability of the Kalisz administration to position New Bedford to receive vast federal and state brownfields resources to support its redevelopment. In describing the former deputy mayor, Larry Silverman, one former city official explained, "Larry's feeling was the best thing was to position ourselves for brownfields." Another former official said: "this concept of brownfields was being born . . . There was a sense of urgency to seize the opportunities, create opportunities."

The City of New Bedford saw brownfields as their ticket to harnessing external resources and position the city at the cutting edge of a newly popular development movement. With the emergence of the brownfields idea, niche developers have grown and focused their energies on the kind of real estate that New Bedford is rich in: unwanted, polluted, and dangerous (Meyer and Lyons 2000). While little in the wider brownfields movement targets HI-TOADS, there is a high level of overlap between the two concepts and the successes the city has had in the HI-TOADS arena can be attributed partially to the city's acclaimed brownfields strategy.

After maritime industries, New Bedford has targeted tourism as its biggest economic development initiative. With its rich history, museums, national park, and miles of coastline, tourism appears to be an easy fit. Central to such an effort is historic preservation. The city's trail-blazing application of urban-renewal funds to adaptive reuse and its later designation as a National Park has created today a well-preserved historic downtown (McCabe and Thomas 1995). But the city's efforts over the last decade to address HI-TOADS has not included any historic preservation elements.

That reality came to a head several years ago when the city spearheaded an effort to redevelop the Fairhaven Mills. The site included several older, derelict structures, one of which was Mill Building #4. These mills sit at the gateway to the city. Their emotional tie to the city's residents was stronger than city leaders imagined in their plans to demolish the site and build a Home Depot. "Growing up as a kid . . . anytime you go on I-195, you always remember seeing those mill buildings over there . . . It's etched in people's minds, if it's not there, it would be a piece missing in the landscape." This historic preservation conflict eventually led to the dissolution of the Home

Depot plans and the HI-TOAD site remains idle today. But the city's efforts at Fairhaven Mills were very proactive, as opposed to how it addressed HI-TOADS in the rest of the city—reactively. Some city officials had no patience for the preservationist: "Well, the mills, those are dirty old places . . . people wanted to transform mills into museums, I mean this isn't Mass-MOCA [the Massachusetts Museum of Contemporary Art]," commented one former official. Most were frustrated at the process by which sites come into the city's hands.

In front of the Elco Dress Factory, a leading city official expressed his concerns: "structurally it's beyond saving. This is the problem in New Bedford. We get these buildings after they're beyond when they can be saved." After the property owner is no longer in receipt of rents, it becomes difficult for him or her to continue to cover basic maintenance and the building falls into a state of disrepair. Only at the end of a prolonged tax foreclosure process does the city control the property. By that time, the building's structural problems are often too severe to promote adaptive reuse.

Another city official recognized the advantage offered by the granite mill construction in neighboring Fall River, explaining that they have been more successful in promoting a "much wider range of reuse there." The brick mills of New Bedford are more susceptible to structural difficulties and make the adaptive reuse of city-owned HI-TOADS challenging.

In the same way that many of Pittsburgh's HI-TOADS were reused because of the catalytic efforts of NGOs, the City of New Bedford expended very few resources at its HI-TOADS without the forceful demands of NGOs. Ironically, these citizen groups are comprised primarily of local residents with little understanding of the complexities of real estate development and little access to power. Yet, by organizing and fighting for what they believe in, they have been somewhat successful.

Given the complexities of real estate law and environmental contamination, many local activists felt it to be a real challenge to speak proficiently about the conditions at their local HI-TOAD site. "People [from the state and federal government] were coming in and talking to us, we did know when people were bullshitting us . . . The state says air is clean, we tend not to believe it," said one community leader.

At Morse Cutting Tools, an area expert on environmental issues, Buddy Andrade, was invited to work with local community groups and residents. Mr. Andrade is executive director of the Old Bedford Village, a local nonprofit development organization. "Buddy Andrade, he really did a lot of work, really opened up our eyes to the environmental issues, let us know what this chemical was, what that was," explained a South Central neighborhood activist.

According to a former public official, the city, working with its partners at the state and federal government, sponsored

> an initiative in the city to educate members of the community on brownfields development and reuse . . . become empowered to facilitate efforts in their own corners of the city. Non-site specific effort, we recognized that people who live in proximity to HI-TOADS are also the people with limited access to healthcare, it's hard to bear out the causality, they have higher rates of hypertension, diabetes . . .

The city truly realized the aims of both community empowerment and healthy environment policies with this initiative. The result was further capacity among the NGO sector to address HI-TOADS and a reduction of risks to public health and the environment.

After gaining a clearer understanding of the complexities of HI-TOAD site redevelopment, these community groups have been able to bring the dangers and threats of their local HI-TOAD site to the attention of key political leaders. One community leader explained: "after fighting the city, they finally gave in." And in the case of Morse, one witness to the events reported, "were it not for the people saying it has to be done, it would not have happened for many years."

These experiences also have helped some activists and NGOs grow and increase their capacity. A keener understanding of the process of fighting for one's own neighborhood helps empower people to continue to fight. In the voice of one community leader: "We know that if we want something it's a fight. I find it kind of challenging!"

The New Bedford case study offered a unique contrast to Pittsburgh and Trenton. In this conclusion, I will summarize the key points of the New Bedford case study while briefly comparing the findings with the two other case studies. In chapter 10, I will go further and fully tie the five case studies together in a final synthesis.

In a city administration renowned for a top-down, action-oriented approach to development, the record of HI-TOADS reuse in New Bedford is an unlikely contrast. Despite citywide policies to prioritize idle properties for job growth, neighborhood revitalization, and transportation-related uses, this research shows that the city's actual action at HI-TOADS reflected much more of a sensitivity to neighborhood needs.

. While previous chapters highlighted Pittsburgh's weak market and Trenton's lack of a market as important drivers in each city's HI-TOADS policies and practices, in many ways it is New Bedford's vital bottom-feeder market that provides demand for scores of obsolete mill buildings throughout the city. These bottom-feeders are seen as the last stage in a building's productive use. Once the bottom-feeders leave, then a property owner often no longer pays the taxes or performs basic maintenance on a property. During the tax foreclosure process, the building tends to suffer structural damage due to neglect or abuse and is suitable only for demolition once it arrives in the city's inventory.

In New Bedford, the bottom-feeders have served as a rudimentary form of protection for these buildings and their rental payments have paid for some building maintenance (if not a high level). Of the five HI-TOADS in New Bedford, three were occupied by bottom-feeders prior to their ultimate abandonment (Pierce Mills, Elco Dress Factory, and Fairhaven Mills), while the other two were so heavily contaminated that once their major occupier left they were essentially fenced off to any use (Aerovox and Morse Cutting Tools).

The federal role in New Bedford's HI-TOADS was one of the reasons that I selected the city for a case study. While the federal government has retrenched its support for urban redevelopment in recent decades, brownfields programs at several federal agencies have begun to reverse that trend by pouring attention and resources into cities to help them address their moribund properties. The result in New Bedford was three-fold: (1) the presence of a federal employee in City Hall actively contributing to citywide brownfields and redevelopment policies, procedures, and education; (2) a level of energy and pride generated by the selection of the city as an EPA Brownfields Showcase Community; and (3) the assembly of a broad coalition of local, state, and federal entities focused on the brownfields agenda brought a wide array of financial and other resources to support HI-TOADS reuse.

EPA also named the City of Trenton a Brownfields Showcase Community and assigned one its own employees to work in City Hall. But the federal government was seen as only a minor partner. In New Bedford, the involvement of Mayor Kalisz on the national brownfields scene and the interest by his political appointees in putting the city at the forefront of a major movement made the federal government a major partner, benefactor, and supporter. City officials built on that support to garner other external resources and to improve the New Bedford market.[6]

[8]

Planning for a Shrinking City

YOUNGSTOWN, OHIO

Youngstown, Ohio, is perhaps most famous for its steel mills, labor disputes that occurred there, the tragedy of their closure, and the unusual tactics taken by workers and religious leaders to resurrect them. In this chapter, I look over the last decade to see what is happening in Youngstown. Through an examination of HI-TOADS reuse, I will probe the lingering trauma of "Black Monday" and the prospects for the future of this shrinking city.

A former political leader of Youngstown spoke about how things were when the steel mills closed. "There was a lot of environmental problems. Kids were going down and fishing in the creek. Young kids were playing in those areas, they were exposed to a lot of hazardous materials. My own son was going down there, I went totally crazy!" That craziness spurred many in the city's government to take quite bold action to clean up and redevelop those former mill sites. What follows is the story of Youngstown's response.

Key Questions for Youngstown

The City of Youngstown received significant attention with the introduction in 2005 of its Youngstown 2010 Citywide Plan, which called for a smaller, better Youngstown. The *New York Times Magazine* listed Youngstown's efforts at creative shrinkage as one of the best ideas of the year (Lanks 2006). How does this innovative planning approach connect with the city's efforts to address HI-TOADS?

A second question in this case study stems from the city's unique history in plant closings. Unlike in other communities hit hard from plant closings in the 1970s, Youngstown's active religious community took a lead role in developing a response. Given that history, how does the religious community in Youngstown become involved in HI-TOADS reuse projects?

With a population of only 3,000 on the eve of the Civil War, Youngstown exploded to 45,000 by 1900 and peaked at 167,000 in 1940 (Buss and Redburn 1983). The city's rapid growth was fueled by an ever-expanding steel industry that by the middle of the twentieth century "had come to dominate every aspect of life" (Buss and Redburn 1983, 2). But the region's growth began to reverse in the 1950s and by the 1970s the seeds were sown for the collapse of the steel industry (Fuechtmann 1989).

September 19, 1977, is known in the Mahoning Valley as Black Monday. It was the day that Youngstown Sheet and Tube Corporation announced the closure of their Campbell Works Plant and the layoffs of 4,100 workers (Kamara 1983). Black Monday was followed by a sequence of other closings that in under three years eliminated over 10,000 jobs from the region (Buss and Redburn 1983). A leading Youngstown researcher said that "when mills closed, neighborhoods closed."

A group of religious leaders banded together in the face of such calamity and formed the Ecumenical Coalition. The coalition attempted to purchase the closed mills and operate them as workers' cooperatives. The effort received much popular and academic acclaim, but ultimately failed (Fuechtmann 1989; Safford 2004). The local coalition requested federal support for the initiative and the request was rejected. But this expression of community interest and involvement is highly unusual and its continued role in the building of Youngstown's future is worth exploring.

From 1970 to 2000, Youngstown's population plummeted 42 percent from 141,000 to 82,000. The impact of the steel industry's closure has been devastating. The area's residents faced a wide range of traumas related to unemployment, but some sociologists argue that the challenges run even deeper. Robert Bellah's (1985) landmark sociological study of individualism and communities introduced the idea of "communities of memory." That is, our communities are constructed based on how we, together, remember the past. In her 2002 study of work and memory in Youngstown, Sherry Lee Linkon argues that the Youngstown community members' memory of themselves is "constituted by their past" and that "until Youngstown addresses its internal conflicts over how to remember its past, it will continue to falter in its efforts to build a new identity for the future" (5).

HI-TOADS in Youngstown

Five HI-TOADS have been identified in Youngstown: an abandoned steel mill at 229 East Front Street (Chevrolet Center), the former Republic Hose site (Aeroquip), the Building and Materials Supply Center (YBM), the Ohio Works Industrial Park, and the former Youngstown Sheet and Tube Campbell Works Plant and Republic Steel (YST/RS). See table 8.1 and figure 8.1 for background information and a map locating each of the final five HI-TOADS included in the analysis.

In this section of the chapter, I will describe three of the HI-TOADS—Aeroquip, 229 East Front Street, and YST/RS—provide background on reuse and redevelopment efforts, and present the results of statistical analysis of census data for each site (see table 8.2). I will then synthesize the data analysis for all three HI-TOADS and present a summary of what I found with respect to HI-TOADS redevelopment, socio-economic changes, and property values in Youngstown.

Aeroquip

Up until 1979, the Aeroquip Corporation manufactured hydraulic hoses at a 48-acre facility along Crab Creek, just north of downtown Youngstown. The plant's closure was covered by *Time Magazine* because its employees bought the complex from Aeroquip and decided to run it themselves ("Buying Jobs" 1979). This bold action was a logical next step in a growing, Youngstown-based movement for cooperative ownership of industry. However, the worker-owner model failed rather quickly at the Aeroquip site and almost three decades later the property remains idle.

Perhaps "idle" is not descriptive enough. "It truly looks like Beirut," commented one city official (see figures 8.2 and 8.3). The Aeroquip site is a frightening stage for half-demolished, rotting industrial waste. The site is so scary that Hollywood filmmakers have descended on the property for an array of films about death, destruction, and horror (Brahler 2005). A series of recent fires have left the remaining structures charred and dangerous.

Shortly after the site was abandoned, looters took all the available valuable material. City authorities suspect that the looters also disrupted electrical

Table 8.1

Summary Information for HI-TOADS in Youngstown

Site name	Size	Location	Historical use	Proposed use	Status of redevelopment efforts (as of 2006)
Aeroquip Site	8.5 acres	Albert Street	Hose manufacturing	Industrial	City is assessing environmental conditions of the site.
YST/RS	1,470 acres	Between Poland Avenue and Wilson Avenue	Steel mill complex	Industrial	Portions of the site have been redeveloped.
Ohio Works	135 acres	Steelton neighborhood	Steel mill	Industrial	Portions of the site have been redeveloped.
229 Front Street	26 acres	East Front Street	Steel mill	Convocation center	Site was redeveloped in 2005.
YBM Site	3.5 acres	Logan & Hubbard Avenues	Building and supply operation	No plans in place	Remediation is underway.

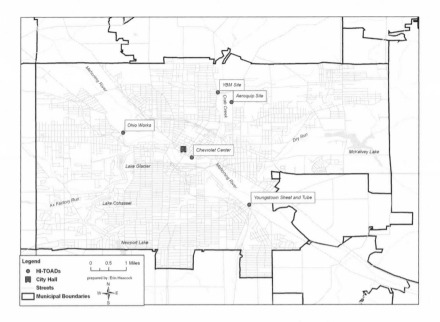

FIGURE 8.1. HI-TOADS in Youngstown, Ohio. Prepared by Erin Heacock.

transformers and introduced a severe PCB contamination problem. The City of Youngstown acquired the site through tax foreclosure and between 1985 and 1992 spent $3 million on environmental remediation and demolition (Smith 2002). Despite those expenditures, the site has remained contaminated. It was listed on the U.S. EPA's Superfund National Priority List, but was never funded. Since 2002, the city has been actively seeking further funding from the State of Ohio through their "Clean Ohio" program (Smith 2002). In July 2006, the state awarded the city a Clean Ohio grant in the amount of $289,830 to complete a Phase II Environmental Assessment.

The City of Youngstown employed both healthy environment and economic development policies in addressing Aeroquip. The site's contamination and public safety challenges so far have been the most important drivers for city action. But the Youngstown 2010 plan calls for a green industrial future land use of the site. Aeroquip is part of a preliminary vision developed by the city and Youngstown State University for an industrial greenway from the school, into downtown, and northward along Crab Creek. The greenway would feature bicycle paths, walking trails, and "green"

Table 8.2

Property Value and Demographic Data for HI-TOADS and the City of Youngstown, 1980 to 2000

	1980	1990	2000	Percent change		
				1980–1990	1990–2000	1980–2000
City of Youngstown						
Mean housing value ($)	15,423.63	16,667.01	21,462.63	8	29	39
Mean gross rent ($)	220.00	319.00	414.50	45	30	88
Percent rental units	0.31	0.32	0.31	4	–3	1
Percent vacant/other units	0.03	0.04	0.08	46	74	154
Average household income ($)	15,435.16	20,281.52	30,149.33	31	49	95
Population	115,579	95,937	82,026	–17	–15	–29
Percent African American	0.33	0.38	0.45	14	19	35
Percent college graduates	0.07	0.08	0.10	14	17	34
Aeroquip Site						
Mean housing value ($)	7,502.09	7,953.54	8,803.76	6	11	17
Mean gross rent ($)	94.00	143.00	273.00	52	91	190
Percent rental units	0.50	0.53	0.50	7	–6	0
Percent vacant/other units	0.02	0.02	0.09	13	269	316
Average household income ($)	11,005.04	15,102.69	23,688.08	37	57	115
Population	4,589	3,797	3,899	–17	3	–15
Percent African American	0.65	0.74	0.80	14	7	22
Percent college graduates	0.02	0.03	0.03	73	–14	49

Table 8.2 (*continued*)

	1980	1990	2000	Percent change		
				1980–1990	*1990–2000*	*1980–2000*
YST/RS						
Mean housing value ($)	19,277.44	20,428.84	27,253.72	6	33	41
Mean gross rent ($)	236.00	352.50	487.50	49	38	107
Percent rental units	0.15	0.21	0.22	36	6	45
Percent vacant/other units	0.03	0.01	0.06	-54	332	100
Average household income ($)	17,245.15	22,393.07	32,674.18	30	46	89
Population	6,386	5,540	4,834	-13	-13	-24
Percent African American	0.06	0.07	0.21	14	179	218
Percent college graduates	0.04	0.05	0.07	11	56	73
Ohio Works						
Mean housing value ($)	10,370.45	9,386.48	14,221.92	-9	52	37
Mean gross rent ($)	182.00	280.00	479.00	54	71	163
Percent rental units	0.34	0.31	0.31	-10	2	-8
Percent vacant/other units	0.02	0.06	0.05	197	-14	157
Average household income ($)	14,841.34	17,294.43	25,440.90	17	47	71
Population	1,396	1,253	1,128	-10	-10	-19
Percent African American	0.14	0.11	0.09	-21	-14	-32
Percent college graduates	0.03	0.08	0.01	179	-83	-52

229 Front Street

Mean housing value ($)	972.58	491.94	n/a	-49	n/a	n/a
Mean gross rent ($)	64.00	131.00	143.00	105	9	123
Percent rental units	0.79	0.84	0.74	7	-12	-6
Percent vacant/other units	0.02	0.03	0.17	26	567	741
Average household income ($)	6,315.60	5,916.63	8,566.15	-6	45	36
Population	778	1,018	1,039	31	2	34
Percent African American	0.30	0.43	0.63	47	45	113
Percent college graduates	0.09	0.05	0.03	-42	-36	-63

YBM Site

Mean housing value ($)	8,648.40	6,703.08	8,772.81	-22	31	1
Mean gross rent ($)	199.50	306.50	383.00	54	25	92
Percent rental units	0.53	0.52	0.46	-3	-10	-13
Percent vacant/other units	0.02	0.06	0.15	174	140	558
Average household income ($)	13,325.42	19,256.27	29,646.74	45	54	122
Population	7,493	6,281	5,067	-16	-19	-32
Percent African American	0.19	0.44	0.49	125	11	150
Percent college graduates	0.14	0.12	0.09	-16	-26	-37

industrial activities. The economic growth along the greenway would come largely from YSU spin-off companies comprised of young entrepreneurs who like to bike to work along a creek. While only in an embryonic stage, the city is committed to cleaning up the Aeroquip site and preserving an industrial use at the property.

In table 8.2, I present census data for 1980 through 2000 for Aeroquip's host census tract. The largely poor, African-American neighborhood has experienced a loss of 15 percent of its population from 1980 to 2000. City action at Aeroquip has happened independently, with little public participation and no involvement of nongovernment organizations.

The overwhelming expenses related to remediating the Aeroquip site has forced the city to rely on external state and federal funding. Almost three decades after the complex closed, the waiting continues. Meanwhile, Aeroquip remains a negative drag on its surrounding neighborhood. One city employee expressed his commitment to turning the site around "before I retire," explaining that he grew up in the neighborhood and feels that cleaning it up is important.

FIGURE 8.2. Aeroquip Site: View of abandoned structures looking north.

FIGURE 8.3. Aeroquip Site: View of abandoned structures looking west.

229 East Front Street

Wedged between the Market Street Bridge and the South Avenue Bridge, 229 East Front Street had for decades been an ideal location for a steel mill. The mill's closure in the 1980s was followed by building demolition and clearance in the 1990s. The 26-acre site stands out from other HI-TOADS in this study due to its downtown location—the site is actually within Youngstown's Central Business District. In most American cities, a prime, waterfront, unimproved piece of land within a CBD would not remain fallow for long. In Youngstown, it took about a decade for the 229 East Front Street site to be reused, a delay that can be attributed to low market demand in the area and the challenges of brownfields' development.

In 2000, a local businessman, Bruce Zolden decided to pursue the expansion of an existing arena he owned in nearby suburban Boardman. Mr. Zolden used his ties in Ohio politics to secure a $1.5 million grant from the State of Ohio for preliminary expenses related to site acquisition and clearance. Then, he approached his congressman, James Traficant, to see how the federal government might be able to help. Representative Traficant suggested that instead of an expansion in Boardman, the project take the form of a new stand-alone community convocation center in the City of Youngstown. With this new form, the congressman felt that he could bring in upwards of $30 million in federal funds. Representative Traficant later would be convicted of bribery, racketeering, and fraud and sentenced to eight years in federal prison. But while in office, Mr. Zolden and Representative Traficant worked together to advance the project. Mr. Zolden contributed several hundred thousand dollars and in-kind services, while Representative Traficant obtained the remainder of the money through the federal coffers.

The City Council and Representative Traficant selected the East Front Street site for their arena. The entire site-selection process was conducted with only a single City Council session devoted to public participation. The Chevrolet Company paid for naming rights and the Chevrolet Center opened for business in 2006 (see figure 8.4). For the purposes of this study, the site was no longer a HI-TOAD by 2002, as heavy equipment began to appear on-site and it took on the appearance of a construction site.

In most U.S. cities, the centrally located 26 acres along the Mahoning River would be a hot commodity. In Youngstown, it sat vacant until an alliance of political activists decided to build a $30 million convocation center there. The city's efforts at 229 East Front Street were driven by economic development interests, in their quest to site a new federally funded construction project.

A look at census data reveals a modest increase in rental values from 1990 to 2000, during the period leading up to the final deal for the convocation center (see table 8.2). Housing value data is unavailable for the host census tract in 2000 (for unknown reasons), but any real improvements in values would not be expected to materialize until closer to 2002.[1] Overall, the census data offers little evidence about how the surrounding neighborhood changed.

During my visit to Youngstown, I explored the area surrounding the completed Chevrolet Center and found few examples of spin-off benefits to the surrounding area. No new restaurants or housing had taken advantage of the new project. However, I had never before visited Youngstown, so it was hard for me to assess what the area was like before the construction began at the site—that is where we can learn from the census data. Before construction began, the surrounding area had a relatively high percentage of elderly residents, a low level of college graduates (3 percent in

FIGURE 8.4. 229 East Front Street: View of recently constructed Chevrolet Center.

comparison with a citywide average of 10 percent in 2000), a high vacancy level—more than twice the citywide levels—and an average household income of just $8,566.

Youngstown Sheet and Tube and Republic Steel (YST/RS)

The massive layoffs at the Youngstown Sheet and Tube's Campbell Works site began in 1977 on what is known as Black Monday and eventually left the entire Mahoning County with a landscape of abandoned steel mills. By far the biggest devastation was in Youngstown, at the Campbell Works site between Poland and Wilson avenues in the city's South Side. The steel mill closed in 1977 and its next-door neighbor, Republic Steel, closed its doors just four years later. Combined, the two mills covered 1,470 acres across Youngstown and neighboring Struthers and Campbell (see figure 8.5).

Residents found it difficult to accept this wholesale shutdown of industrial activity. Locals were used to the cyclical nature of steel; many expected that the hiring would restart shortly. Those 14,000 lost jobs were never replaced. Instead, the mills began to crumble from non-use.

FIGURE 8.5. Youngstown Sheet and Tube/Republic Steel: Looking south toward the City of Struthers, mill ruins on a portion of the site awaiting redevelopment.

In 1991, Congressman Traficant again stepped in as a key figure in HI-TOADS redevelopment. Along with business and academic leaders, he formed the Mahoning River Redevelopment Project, with the purpose of developing an inventory of brownfields in the region (YSU 2001). In 1995, a group of politicians joined together with those involved in the redevelopment project to create a new entity, the Mahoning River Corridor of Opportunity (MRCO), to plan strategically for the entire 1,470-acre multi-jurisdictional site (Wheatley 2000).

MRCO is led by the mayors of the three host cities, Campbell, Struthers, and Youngstown. Several major business and economic development organizations are also involved, as is the Eastgate Regional Council of Governments, Mahoning County, Youngstown State University, and the area's major utility companies.

At a time when critics were calling for the Youngstown community to come together, this was a truly forward-thinking approach. The group's early efforts were recognized in 1999 at a White House Joint Center for Sustainable Community Awards, sponsored by the United States Conference of Mayors and the National Association of Counties. Then, in 2002, MRCO released its master plan calling for light industry, office and commercial services, outdoor storage and general manufacturing, and open space (see figure 8.6).

Concurrently with the planning process, MRCO officials recognized the urgent need for a new bridge to facilitate truck access to wide interior swaths of the site. Through a web of federal and state funding, the bridge was completed in 2005. The span crosses all three cities in the MRCO and was dubbed the Steelworkers Bridge to commemorate those workers who helped to build the region (see figure 8.7). It serves as a metaphor for what regional cooperation can accomplish, and in the case of a HI-TOAD site, what is necessary to accomplish.

In addition to the MRCO efforts, the City of Youngstown developed Performance Park, an industrial park, in a small section of the site in the early 1990s. MRCO has had some on-the-ground success largely in Campbell and Struthers, while most of the 1,470 acres remain idle. With the construction of the new bridge in 2005, MRCO is quite optimistic about the prospects for redeveloping the remainder of the site for industrial uses.

FIGURE 8.6. Master plan for the reuse of the YST/RS site. Eastgate Regional Council of Governments.

FIGURE 8.7. YST/RS: Steelworkers Bridge, dedicated in 2006.

MRCO was an active player in the City of Youngstown's 2010 Plan and MRCO's land-use plan for the site even appears on page 83 of the city's plan. While master plans examined in this study rarely referenced HI-TOADS, the Youngstown 2010 Plan refers specifically to the MRCO planning effort in its recommendations.

YST/RS became a HI-TOAD site after 1980, once all of the jobs left. For decades, it has been a drag on its surrounding neighborhoods, but that is beginning to change. It has not completely changed—much of YST/RS is vacant or "improved" by abandoned buildings. While the MRCO efforts are laudable, interviews with local officials revealed that the portions of YST/RS within the boundaries of Youngstown remain a HI-TOAD site.

The City of Youngstown's participation in MRCO speaks to its commitment to economic development approaches at YST/RS. The diverse membership within MRCO includes public agencies, economic development agencies, business groups, utilities, and a state university—but there is no direct role for community-based organizations. The City of Youngstown's involvement is organized around a unitary interest: bringing new employment to the YST/RS site. The city did not draw on healthy environment or community empowerment approaches in this case.

An examination of census data is particularly challenging for the YST/RS site, because it covers two census tracts and three municipal governments. Population in the host tracts fell by a quarter from 1980 to 2000, but other social and wealth indicators remained constant or improved. In particular, both housing value and gross rents increased at rates faster than the city as a whole. MRCO's local and national acclaim may have begun to change area sentiment, even though much of the site continues to have a negative drag on area property values. Major events like the construction of a new bridge and the recruitment of new industries to the park (even those outside the City of Youngstown) may be having a mitigating impact on the area's overall decline.

While not a focus of this research, Youngstown's late-1980s effort at Performance Place (the small, successfully redeveloped industrial park at the YST/RS site) also suggests a level of hope and optimism for the remainder of the YST/RS site. Performance Place is a fully built-out industrial park within the City of Youngstown, on former YST/RS property. Everybody included in this study of Youngstown wanted to brag about the success of Performance Place and hold it up as a model of what can happen at other HI-TOADS in Youngstown. Between that success story and the potency of the MRCO planning effort, the prospects are palpable.

How Are HI-TOADS Addressed in Youngstown?

The City of Youngstown has received national acclaim for its Youngstown 2010 master plan. Unlike other comparable cities, Youngstown has accepted its population decline and through the plan is attempting to make this smaller city "cleaner, greener, and more efficient" (City of Youngstown 2005, 7). The plan includes a thorough inventory of all vacant land and highlights all significant sites (including references to each HI-TOAD site).

City officials explained that the next step in the master planning process is to develop detailed neighborhood plans where the need exists. Ostensibly this effort would focus around each active HI-TOAD site.

Within the last ten years, the five HI-TOADS in Youngstown have attracted quite vigorous city interest. One site, 229 East Front Street, was redeveloped completely by the city, with the help of state, private, and federal funding. Two others, YST/RS and the Ohio Works, have been reused partially thorough active economic development efforts by the city. And the city is moving forward with environmental work at the final two, Aeroquip and YBM. Overall, the city has taken a major role in HI-TOADS, relative to other cities in the study. One business leader assessed that the city's "cleaning up sites, making them available for development are number one positive thing they have done."

Primarily through the city's economic development agency, the city offers discounted land, tax incentives, and, in some cases, indemnification from future liability associated with industrial contamination. A business leader explained that the city's strategy is fueled by "the use of the city's full faith and credit to indemnify owners." Such indemnification combined with the wide range of tax breaks and grants means that relative to greenfield development in the surrounding suburbs, the "city had put a product [urban land] on the table that's pretty hard to beat," commented another business leader.

Mahoning County has been supportive of the MRCO's efforts at YST/RS, but beyond that site, the county has yielded to the city government at other HI-TOADS. State and federal environmental agencies have funded practically all of the investigation, remediation, and redevelopment at Youngstown's HI-TOADS. State funding came in the form of a bond issue in 2002 that raised $200 million over several years for brownfields redevelopment. While federal brownfields funding often requires a match, the state's Clean Ohio program has an exemption to the match for distressed cities (for which Youngstown qualifies). These Clean Ohio funds helped focus the City of Youngstown's efforts around some of its smaller, yet serious brownfields: "Until Clean Ohio funds came, we didn't think we wanted to clean 10- and 12-acre sites," remarked a city official.

MRCO's receipt of a federal EPA brownfields grant in 2003 was instrumental in initiating an inventory of the region's brownfields. That seed money was then followed by further federal and state aid at both YST/RS and 229 East Front Street. Interestingly, both efforts were led by the incarcerated former congressman James Traficant (in jail due to his bribery and racketeering convictions). The influence of a single federal politician in redeveloping HI-TOADS was unique to Youngstown (in comparison with the other case-study cities). Trraficant's unusual role and subsequent fall is, at the very least, intriguing, and perhaps suggestive of the challenging political, ethical, and financial hurdles that must be overcome in order to reuse HI-TOADS.

MRCO is the most significant NGO in HI-TOADS reuse efforts in Youngstown. As discussed earlier, MRCO is comprised of business, academic, and governmental bodies, with no community group representation. Looking citywide, community groups are essentially absent from HI-TOADS reuse efforts. The neighborhoods abutting both Aeroquip and the YBM site have neighborhood associations, but little activism comes from those groups for addressing HI-TOADS. One local community leader told me, "at this point, I don't think we have a voice in what happens."

In fact, it appears that the city has made a systematic effort to create neighborhood planning districts so as to isolate the Aeroquip site from their abutting residential blocks. A city official remarked to me, "sometimes development and neighborhoods don't match-up—the whole NIMBY thing."

In figure 8.8, I present a detail of the 2010 Plan's neighborhood boundaries. The Aeroquip site sits squarely within Block 3002 (Census Tract 8005), with boundaries outlined as neighborhood number 87. According to the 2000 Census, two people live within neighborhood 87. This tactic of creating neighborhood planning units without residents, a type of gerrymandering, allows the city to exclude HI-TOADS reuse from consideration in abutting residential neighborhood planning exercises.

Beyond disempowered community groups, there are other active NGOs in the Youngstown area, including the regional chamber of commerce and a faith-based organization named ACTION. While deeply interested and involved in redevelopment discussions and promoting Youngstown, these other NGOs are not involved in HI-TOADS reuse. The Ecumenical Coalition played an important early role in fighting the closure of steel mills, but the descendant organization, ACTION, focuses its work more on neighborhood-oriented quality of life issues such as recruiting a new supermarket into an area with none.

The other type of NGO involved in Youngstown HI-TOADS reuse is businesses. In the 229 East Front Street redevelopment, Mr. Zolden played an important role, but largely the business community was absent from

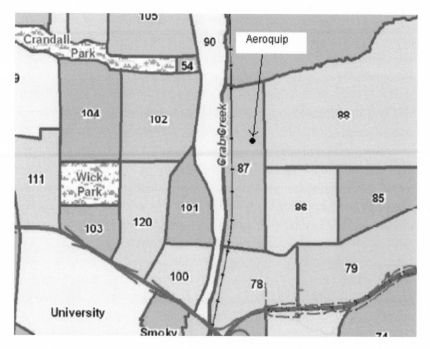

FIGURE 8.8. Neighborhood 87. The Aeroquip site was put into a planning district defined by the city of Youngstown as neighborhood 87, with an estimated population of two people. Youngstown 2010 Plan.

most HI-TOADS initiatives. While it was outside of the scope of this study, US Steel played an important role in HI-TOADS reuse prior to 1995. When their mills in Youngstown began to close, "they had an industrial development team in place, they were involved in getting all of their buildings down and get into the hands of someone else," explained a regional NGO leader. He continued, "they had a branch office in Youngstown, just to oversee the demolition of former steel mill buildings, grading it, and preparing it." Some speculate that "they wanted to do some positive things for the community," but that all ended by the mid-1990s when US Steel had extracted itself fully from Youngstown and other corporate leaders were less community-focused in their closures.

Through formal and informal efforts, the city and MRCO have explicitly identified HI-TOADS and devised strategies for redeveloping them. The

229 East Front Street case is somewhat different in that it was developed in response to a political interest in construction of a highly subsidized arena. The arena simply needed a site and city leaders selected 229 East Front Street because of its size and location within the CBD. For the city's other HI-TOADS, there was (and continues to be) an explicit effort by the city and MRCO to turn the crises of abandonment into an opportunity for new employment.

While the YST/RS site and Ohio Works are only partially redeveloped, local efforts there have been Herculean. The city has overcome long odds to remediate these sites and get them back into productive reuse. The failures at YBM and Aeroquip can be attributed to the higher price tags on remediation at those sites and to their poor location (principally bad highway access).

Key Themes in Youngstown HI-TOADS Redevelopment

Economic development approaches dominated Youngstown's approach to HI-TOADS. In YST/RS, 229 East Front Street, and the Ohio Works site, the city focused its efforts on getting employment- and tax-generating uses at each site. With such a concerted effort, the city was largely successful in doing just that.

The city's failures, thus far, have been at YBM and Aeroquip, where the city has taken more of a healthy environments approach to reuse. At those two sites, the city predominantly directs public policies to cleaning the sites and controlling the public health risks associated with them. While each site is slated for a continuing industrial use, it is worth emphasizing that the plan is to redevelop each site within a "green industrial" zoning classification. This notion of "green industrial" is a variant on the tripartite sustainable development model—environment, economy, and equity/social justice—where industry can be eco-friendly (Campbell 1996). Even though green industry is a powerful concept with great promise, it is quite telling that YBM and Aeroquip remain vacant today.

In many ways, the Youngstown experience with HI-TOADS differs remarkably from the other cities included in this book. While certainly imperfect, the city's innovative 2010 Master Plan and the nationally recognized MRCO planning effort demonstrate a unique commitment to long-term, visionary, and holistic approaches to the reuse and redevelopment of HI-TOADS. As highlighted in the earlier case studies of Pittsburgh and New Bedford, those cities did look to the future in many of the same ways as Youngstown, but neither city involved comprehensive public participation or codified their plans. In Trenton, there was evidence of such formal planning, but those plans largely did not include direct references to HI-TOADS. In Youngstown, they did.

Every active HI-TOAD site in Youngstown was included in the 2010 plan. For each one, some consideration was made about a path forward and such decisions were generated through an extensive citizen-participation process. The city looks to go further: one official said, "our next focus in the 2010 [planning effort] is more detailed neighborhood plans. We would look at buildings like [HI-TOADS]—a more detailed analysis." The city hopes to continue to use formal planning to guide the reuse of HI-TOADS.

Interestingly, two active HI-TOADS in desperate need of reuse have been targeted such that they will be excluded systematically from such detailed neighborhood planning. A city official explained that "some of the neighborhoods were sectioned off so they would be neighborhoods in their own . . . No people live in these 'new' neighborhoods." Through this process, one HI-TOAD site, Aeroquip, was placed into its own "neighborhood" with an abutting industrial park. When I asked about the impact of this "sectioning-off" on citizen participation, one official explained "the residents will be told what the plans are." This troubling strategy has the potential to accelerate the reuse of HI-TOADS, but at the cost of meaningful citizen participation.

As described earlier in the chapter, the MRCO likewise has not been completely open to the public in its dealing, because it is comprised of a board representing organizations, public agencies, and businesses, with no allotted role for community groups or ordinary citizens. Nevertheless, the group's innovative, regional approach has yielded much success (as measured by their external recognition and awards) in an area desperately in need of hope. MRCO is headed by the charismatic Bill DeCiccio. In describing him, a city official said, "Bill is 'Mr. Youngstown' [he's] been through it all." Through Mr. DeCiccio's leadership, MRCO has provided a comprehensive strategy for reusing what may be the biggest and most gnarly HI-TOAD site featured in this book. Never forgetting that fateful Monday in 1977, MRCO stands as a tribute to what was once great about the Mahoning River Valley Corridor and its efforts represent a unique and partially successful route for addressing HI-TOADS.

. . .

The City of Youngstown did not approach its HI-TOADS in a typical way. Perhaps that is because HI-TOADS are part of a broader collapse of nearly every facet of life in this one-industry town. With the sudden and rapid closure of nearly all of the industrial employment in the city, the response over the course of the 1980s, 1990s, and 2000s was driven by sheer desperation.

A former politician remarked: "It was frustrating. At the time, the steel mills killed our General Fund, we had downsizing." With massive unemployment, a rash of social ills, and a dwindling tax base, the city had few options to develop a public policy and planning response to the closed mills. "Out of necessity, Youngstown became very creative in how they approached their demise," said one state official.

Part of that creativity was in how Youngstown handled liability. In the late 1980s, the city began taking on all liability for environmental contamination in exchange for businesses agreeing to reuse former mill sites. This, at a time when the brownfields concept started gaining ground and liability was widely perceived as the chief impediment to reuse. In Youngstown, it was removed as an impediment by the sheer audacity of city leaders like Pat Ungaro, a former assistant principal and football coach who served as Youngstown's mayor from 1984 to 1997. Ungaro took big risks in getting sites back into productive reuse based on the belief that the city needed bold change. In many accounts, his strategy of gaining control of derelict mill sites, demolishing the buildings, and indemnifying future owners set Youngstown on its current course.

Ungaro's legal and financial gambles appeared to have paid off and continue to guide current city policies at HI-TOADS. One community leader commented on the city government's record: "They've picked some prospects with reasonable geometries, they've cleared them and got businesses there." In many ways, they also have a strategy of pursuing the easiest sites—leaving much of YST/RS, Aeroquip, and YBM fallow. Part of what makes the Youngstown story unique is that, despite the fact that heavy industrial activity haunts much of the city's landscape, little contamination remains. With little in the way of contamination, the city's clean-slate approach stems primary from stigma effects. Unfortunately, in order to ascertain that a given site is clean, a developer must expend funds to conduct environmental tests and assessments. In some cases, such costs may exceed the value of the land.

After the steel mills closed, they remained unused for decades. The unknowns associated with contamination and the functional obsolescence of
the structures kept investors away. The antiquated structures were seen as
deterring any reuse and developers and investors were turned off. "While
mills were there, nobody looked at them. Once [the buildings] were gone,
with some incentives, now [investors are] interested," said a community
leader. The city's attention to site control, demolition, and site preparation
was intended to overcome the stigma of heavy industry and to deliver clean
property. They were trying to compete for industrial users with the surrounding suburban and rural areas who could offer prime greenfield sites,
with no contamination to worry about.

Just as the sudden, traumatic collapse of the steel industry shaped the city's
response to HI-TOADS, it likewise affected local attitudes and approaches to
historic preservation. In the clean-slate approach, there is little room to preserve the aesthetically important or the culturally important physical remnants of Youngstown's past. For many, the past was a glorious time. At its
peak, Youngstown was a vibrant, active center full of wealth and pride. Even
its "dirty" side, its industries, are a permanent part of the local psyche. One
community leader fondly recounted to me the thrill "when they poured the
[steel processing] ladles, it was like nothing else, like fireworks." Youngstown's efforts to forge ahead to the future and its strong connection to the
past has been manifested in HI-TOADS reuse not through historic preservation of structures, but rather in the preservation of historic land-use patterns:
the siting of industrial uses in close proximity to residential uses.

The city's clean-slate policies have left few historic industrial buildings
standing.[2] In fact, Youngstown's mayor embarked on an ambitious $1.5 mil-
lion abandoned-building demolition program in 2007. City action is driven
by the assumption that new industrial users require a level, unimproved,
remediated piece of land, indistinguishable from a greenfield. This approach is grounded on a second, equally important assumption: that preserving the historic land-use patterns is a good idea. In most cities developed during the Industrial Revolution, residential areas were built within
walking distance to industrial operations—known today as "workforce
housing." Youngstown is no exception; its steel mills (current and former)
are all surrounded by densely built residential neighborhoods. While other
cities have used the closure of mills as a springboard for eliminating this
"mixed-use" phenomenon, Youngstown took a different tack.

The city proposed a new land-use zone in the 2010 plan: industrial green. This new industrial zone requires a low-impact variety of industrial use, including a prohibition on outdoor storage and strict landscaping requirements. As one city official put it, "they won't have that much of an impact on neighborhoods." In many ways, the "green" in industrial green was a compromise. According to one community leader, "we put the industrial green district to say to our friends in [the city's office of] Economic Development that this is not just about economic development." Another community official added, "this is not going to be industrial like you know it."

In many senses, the need for an industrial green category speaks to the small amount of land in the city available for high-paying employment opportunities. On another level, the city is actively trying to preserve something special about its landscape, in the same way that preservationists try to save individual buildings and districts. By maintaining scores of acres of "industrial green" zoned land in close proximity to residential areas, the city is perpetuating a nineteenth-century land-use pattern. But they do propose to make changes. Their vision for the future Youngstown involves a "greener" industrial base that can co-exist better with recreational, residential, and institutional uses (such as Youngstown State University). One example is the city's vision for redeveloping HI-TOADS in an way that will improve public access to the rivers and creeks. In describing historic public access to the Mahoning River, one city official said, "it's fairly hidden— there are people who have lived here their whole lives and didn't know there is a river in Youngstown . . . access to the river was cut off, reserved for industry. Industry used the river as a sewer." Current city efforts seek to improve access to the river and reduce pollution, not by changing the uses to recreational, residential, or commercial, but by modifying the requirements of industrial. Partly, this is a decision to preserve the city's industrial heritage.

In the late 1990s, the Charles Stewart Mott Foundation commissioned a study to consider the future for Youngstown. The report, *Waiting for the Future: Creating New Possibilities for Youngstown*, used the "public capital" concept to identify Youngstown's challenges (Fitzgerald 1999). For the report authors, public capital in a city is comprised of nine dimensions, ranging from having catalytic organizations to having conscious community discussions. The study found severe deficiencies in Youngstown in each of the nine dimensions and recommended bold action to effectuate change.

This call-to-arms was heard. It is apparent that "2010 was a by-product of working together . . . The Waiting Game identified Youngstown as just waiting, it sparked some serious discussions. [John] Swires read it and said we need a comprehensive plan . . . New president of YSU came in and got everyone focused."[3] Swires was an active city council member representing the 7th Ward. His efforts were successful and the 2010 planning effort was seen by many as helping to galvanize the nongovernmental sector to start to work together.

To understand the city's challenges with public capital, it is important to return to Black Monday. As one area official explained, before Black Monday, "they were never a community that worked together, they were separated—immigrant groups were fractured." Another official went further: "the thing that bound the community together was the mills, after [the mills closed] they all went to their own groups—[the mills] was the glue." But with the 2010 planning effort, things began to change (albeit over twenty years of antagonism later). "These immigrant groups had no tools to work together, that was going on until just a few years ago . . . 2010 [planning effort] was a by-product of working together," said an official in nearby community.

Despite the optimism, overall I found little meaningful involvement of NGOs in the reuse of HI-TOADS. One city official explained, "there are not that many community organizations. Those that do exist, they are concerned with their areas, their neighborhoods . . . We are trying to build their capacity, get them involved in planning taking place." Yet it appears that the city is attempting to reduce the capacity of community-based organizations to be involved in HI-TOADS reuse. For example, the formation of neighborhood planning boundaries around the Aeroquip site will have the effect of excluding citizen involvement with reuse planning.

Without a vital community-oriented sector, most cities need to rely only on governmental and business interests. Unfortunately for Youngstown, business leadership is also lacking. "When the mills shut down, there was such a leadership loss. All the corporate leaders left, all the banks were absorbed by larger institutions. We just lost our corporate structure. When you had the corporate power, they kept corruption in check—there was a big vacuum," explained an NGO leader in the valley. A city official who stayed quipped, "a lot of my good buddies from high school live in Florida." A former political leader who worked hard in the 1980s and 1990s to turn Youngstown around was pretty much on his own. "We had no corporate support to do anything."

While there is some hope for a new kind of leadership and growth of public capital, I did not find any evidence that such potential had been realized in the reuse of any HI-TOADS. The story about the development of the 229 East Front Street Site is emblematic of how things get done in Youngstown: by a loose coalition between politicians and a single developer. While Mr. Zolden's efforts were admirable, they were not enough to compensate for the widespread lack of true capacity for action in this shrinking city.

What can urban practitioners and scholars learn from the Youngstown experience? I was first attracted to studying Youngstown because of its innovative planning endeavors. I was hopeful about how those attitudes might be translated into redevelopment action. I was also drawn to studying Youngstown for its unique history as a hotbed of faith-based activism in plant closings.

I found some answers. The faith-based activism so richly documented in the late 1970s and early 1980s is but a shadow of what it once was. The only organization active in redevelopment in Youngstown is ACTION, and they have focused their efforts only on small-scale redevelopment and quality-of-life improvements. Their role pales in comparison to the high-profile involvement of NGOs in Pittsburgh, New Bedford, and even Trenton. The Mott Foundation's *Waiting for the Future* report was correct in 1999 with its assessment that Youngstown lacked the public capital to create a better future for itself. This chapter shows that such public capital is still lacking. The dearth of corporate and philanthropic organizations in the city makes such growth in public capital quite challenging. Through the lens of Youngstown's most serious abandoned properties, the government was the only viable force at work for positive change.

While other cities have high property values and an active development community, Youngstown had neither and any redevelopment has to be done by the city government. One business leader said to me that the "city did what private developers did in Pittsburgh, because [developers] could make a profit . . . Real estate values are quite low [in Youngstown]." With low property values and poor public capital, the city turned to a new paradigm in urban planning: the shrinking city. This idea for a smaller, better Youngstown flows from the work of Popper and Popper, who called for a new kind of planning:

> What can planners do when communities are shrinking rather than growing? We suggest the answer is "smart decline." Smart decline means leaving behind assumptions of growth and finding alternatives to it . . . Planning for less—fewer people, fewer buildings, fewer land uses—demands its own distinct approach. (Popper and Popper 2002, 21)

The 2010 plan does explicitly recognize the need to address HI-TOADS in the city and proposes industrial green zoning for most of them. But it is too early to tell how the 2010 plan will be implemented and if smart decline can truly work in Youngstown. The city's commitment to a historic land-use pattern that brings industry in close contact with residential uses is being framed through a healthy environments discourse, yet is anything but. The new industrial green zoning is "less" noxious than traditional industry, but risks to nearby residential populations continue. In a smaller, better city, are there not better locations for the placement of industry?

[9]

Race, Preservation, and Redevelopment

RICHMOND, VIRGINIA

This final case study explores redevelopment practices in a great southern capital, Richmond, Virginia. With its checkered history of civil rights, preservation, and urban renewal, Richmond is a rich site for exploring the local political responses to HI-TOADS.

In 1996, locals erected a statue of Athur Ashe along Monument Avenue among the statues of five Confederate War heroes. In telling the story of the statue, Robert Hodder (1999) wrote about the "rocky road that must be traveled to advance the social meanings of urban heritage" in Richmond and elsewhere. The road is particularly rocky in Richmond, Virginia, and in this chapter, I show how advancing "social meanings of urban heritage" is but one of the challenging but achievable aims undertaken by Richmonders at six of the city's HI-TOADS.

Key Questions for Richmond

As in the other case studies, I answer two broad questions in this chapter: (1) of those cities that acknowledge HI-TOADS as a problem, what policies do they use to address them? And (2) how successful are those policies? Within the first research question, I am attempting to categorize each cities' efforts into one of three classes of policies: economic development, community empowerment, or healthy environment. After conducting a review of the literature on urban development in Richmond and interviewing city officials in Stage 2 of the research, two narrower and interrelated questions emerged for exploration in this case study.

While not entirely unique in the United States, Richmond was once dubbed the "separate city"—a place where whites and African Americans live apart and a place of great racial strife (Silver and Moeser 1995). While such separateness is evident in cities throughout the United States, racial tension has received a great deal of scholarly treatment in Richmond and is seen by some as a particularly strong example of a segregated city (Silver and Moeser 1995).

Research on Richmond's history has identified race as a central and organizing variable in understanding the city's physical and political landscape. To what extent, then, does the city's past and current racial tensions and segregation affect the redevelopment of HI-TOADS?

A closely tied question is how the city's legacy of urban renewal plays into current redevelopment practices in the city. Urban renewal in the 1950s and 1960s was a tool used by the white political elite in Richmond to maintain their hold on power and to repress the city's African-American population (Byng 1992; Silver and Moeser 1995). As occurred in scores of other U.S. cities, Richmond officials made a concerted effort to destroy the physical remnants of scores of primarily African-American communities. What are the lingering effects of urban renewal on HI-TOADS projects?

First settled in 1607 by English explorers, Richmond quickly took its place among the great southern cities. The city's strategic location along the James River made it a major transportation center early on. By the late seventeenth century, Richmond was a hub of flour mills and tobacco factories (Hoffman 2004). By the nineteenth century, iron and metal works also appeared as major industries. But, as opposed to some of the mono-industry cities explored in this book (such as New Bedford and Youngstown), "numerous other industries provided depth and breath to the local economy" (Hoffman 2004, 9).

Much of Richmond's original growth was fed by the institution of slavery. Slavery was instrumental to the tending of vast farms and the success of the agriculture of the region. During the Civil War, Richmond was the capital of the Confederacy. To the extent that a cultural affinity for the spirit of the Confederacy endures today, Richmond remains that movement's capital. The city's wealth and prominence is tied to slavery, as are its deep and pernicious racial divisions and conflicts.

After World War II, white flight to the suburbs accelerated and the neighborhoods left behind became more and more integrated. Fearful that they would lose power to the growing African-American population, white politicians took two approaches: They expanded the city's boundaries through annexation, and they converted African-American neighborhoods into commercial districts through urban renewal (Silver and Moeser 1995).

The annexation strategy preserved white power for some time, but ultimately expansion could not keep up with outmigration. "Less people resided in Richmond in 1980 than did in the early 1940s, despite a one-third increase in city territory" (Silver 1983, 38). Urban renewal cleared over 1,000 acres and demolished 4,700 housing units, while the white majority continued to shrink (Silver 1984, 1983). Urban renewal planners "refused . . . to acknowledge the role of public policy in the process of neighborhood change itself. They assumed incorrectly that neighborhood decline was merely part of the natural metamorphosis of cities" (Silver 1984, 101).

During this heated time of evictions and bulldozing, the Supreme Court ruled in the 1954 landmark case *Brown v. Board of Education* that separate but equal is inherently unequal. This lasting provision of constitutional law required the *very* separate city of Richmond to make some adjustments. White leaders resisted the *Brown* decision and failed to integrate city schools. Nine years after the decision, African-American children made up 58 percent of students in the Richmond public schools but white schools had only 2 percent African-American students. In African-American schools, there were no white children (Silver and Moeser 1995).

With the ascent of African-American political power in the 1960s, the tide of urban renewal began to shift in Richmond. By 1974, a major re-orientation shifted city officials toward neighborhood conservation instead of neighborhood demolition. But racial strife persisted and continues today. The "separate city" of the past is now a "separate metropolitan area." Few white enclaves remain within Richmond, while the surrounding suburban areas outside the city are predominantly white. The city faces many of the challenges of other large, segregated cities: rising rates of violent crime, homelessness, poor educational attainment, and public health challenges (Silver and Moeser 1995). And like other cities, Richmond's response was urban renewal. At the height of these federally subsidized renewal efforts, "the planner's cure for neighborhood ills sometimes proved more deadly than the disease" (Silver 1984). In this chapter, I will explore how much has changed in Richmond, reflecting on its history of racial division and neighborhood demolition.

HI-TOADS in Richmond

Six HI-TOADS met the criteria for the case study: Manchester Lofts Building, Tobacco Row, Tredegar Iron Works, Fulton Gas Works, Richmond

Table 9.1

Summary Information for HI-TOADS in Richmond

Site name	Size	Location	Historical use	Proposed use	Status of redevelopment efforts (as of 2006)
Richmond Memorial Hospital	14.5 acres	Westwood Ave.	Hospital	Housing	Several developers have attempted to redevelop the property without success. The site has remained vacant since 1998.
Richmond Cold Storage	5.7 acres	17th St. and Clay	Storage	Mixed use	The site has remained vacant since 1980s. Current owners are seeking a buyer.
Tobacco Row	15.5 acres	Between Main St. and Carey St. in Shockoe Bottom	Cigarette factories	Housing, retail, and office	Fourteen of the fifteen buildings are in some form of redevelopment or reuse; only the Lucky Strike building remains vacant.
Fulton Gas Works	9 acres	Williamsburgh Rd. in the Fulton neighborhood	Gas works	Unknown	City has attempted unsuccessfully to market the site for private uses.
Tredegar Iron Works	5 ± acres	3215 E. Broad St	Iron works	Museum	Partially renovated for private museum use in the 1980s. In the 1990s, renovations were completed as part of a lease with National Parks Service.
Manchester Lofts building	0.75 acre	815 Porter St.	Bakery, then light industrial	Housing	Converted into 80 condominiums.

Cold Storage, and Richmond Memorial Hospital. See table 9.1 and figure 9.1 for background information and a map locating each of the final six HI-TOADS included in the analysis.

In this section of the chapter, I describe three of the HI-TOADS—Fulton Gas Works, Richmond Memorial Hospital, and Tobacco Row—provide background on reuse and redevelopment efforts, and present the results of statistical analysis of census data for each site (see table 9.2). I then synthesize the data analysis for all six HI-TOADS and present a summary of what I found with respect to HI-TOADS redevelopment, socio-economic changes, and property values in Richmond.

Fulton Gas Works

Fulton Hill was a predominately African-American neighborhood that was destroyed in 1968 with an ambitious urban-renewal scheme (Davis 1988). Nine hundred privately owned homes, one school, five churches, and all of the local businesses were felled by the renewal bulldozer. The only remaining neighborhood structure was the Fulton Gas Works on Williamsburgh

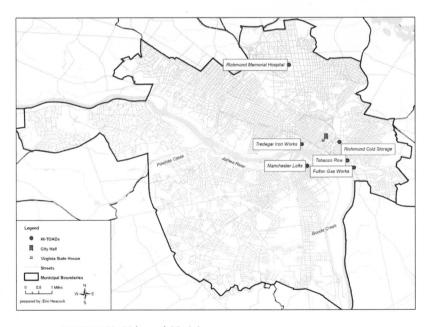

FIGURE 9.1. HI-TOADS in Richmond, Virginia.

Table 9.2

Property Value and Demographic Data for HI-TOADS and the City of Richmond, 1980 to 2000

	1980	1990	2000	Percent change		
				1980–1990	1990–2000	1980–2000
City of Richmond						
Mean housing value ($)	17,515.12	35,177.79	53,427.76	101	52	205
Mean gross rent ($)	227.00	415.00	543.00	83	31	139
Percent rental units	0.50	0.49	0.49	-2	1	0
Percent vacant/other units	0.02	0.03	0.03	40	-3	35
Average household income ($)	17,534.48	31,908.40	45,743.07	82	43	161
Population	218,837	202,644	197,790	-7	-2	-10
Percent African American	0.51	0.55	0.58	8	4	13
Percent college graduates	0.20	0.24	0.29	23	21	49
Richmond Memorial Hospital						
Mean housing value ($)	22,431.50	44,673.51	65,520.45	99	47	192
Mean gross rent ($)	244.00	390.00	460.00	60	18	89
Percent rental units	0.49	0.51	0.51	4	-0.1	3
Percent vacant/other units	0.01	0.01	0.01	32	-20	5
Average household income ($)	19,537.44	42,296.40	50,932.15	116	20	161
Population	6,820	6,826	6,487	0.09	-5	-5
Percent African American	0.25	0.38	0.41	52	9	66
Percent college graduates	0.37	0.39	0.44	5	14	20

Richmond Cold Storage

Mean housing value ($)	4,629.88	14,895.18	18,719.74	222	26	304
Mean gross rent ($)	192.00	358.00	600.00	86	68	213
Percent rental units	0.68	0.66	0.71	-4	8	4
Percent vacant/other units	0.08	0.05	0.04	-41	-20	-53
Average household income ($)	12,278.88	27,024.59	39,809.34	120	47	224
Population	1,885	1,793	2,248	-5	25	19
Percent African American	0.83	0.74	0.54	-10	-28	-35
Percent college graduates	0.16	0.19	0.45	22	135	188

Tobacco Row

Mean housing value ($)	4,629.88	14,895.18	18,719.74	222	26	304
Mean gross rent ($)	192.00	358.00	600.00	86	68	213
Percent rental units	0.68	0.66	0.71	-4	8	4
Percent vacant/other units	0.08	0.05	0.04	-41	-20	-53
Average household income ($)	12,278.88	27,024.59	39,809.34	120	47	224
Population	1,885	1,793	2,248	-5	25	19
Percent African American	0.83	0.74	0.54	-10	-28	-35
Percent college graduates	0.16	0.19	0.45	22	135	188

Table 9.2 (continued)

	1980	1990	2000	Percent change		
				1980–1990	1990–2000	1980–2000
Fulton Gas						
Mean housing value ($)	18,520.49	27,434.37	56,022.29	48	104	202
Mean gross rent ($)	240.00	432.00	553.00	80	28	130
Percent rental units	0.13	0.30	0.32	120	7	136
Percent vacant/other units	0.00	0.03	0.02	—	-38	—
Average household income ($)	14,038.13	22,550.64	51,439.83	61	128	266
Population	729	1,015	1,439	39	42	97
Percent African American	0.52	0.75	0.90	45	19	73
Percent college graduates	0.04	0.15	0.15	321	-1	317
Tredegar Iron						
Mean housing value ($)	n/a	n/a	648.71	n/a	n/a	n/a
Mean gross rent ($)	407.00	967.00	613.00	138	-37	51
Percent rental units	0.80	0.82	0.90	3	10	13
Percent vacant/other units	0.05	0.01	0.03	-70	101	-40
Average household income ($)	12,500.59	24,112.83	30,767.00	93	28	146
Population	2,818	2,229	2,401	-21	8	-15
Percent African American	0.41	0.50	0.45	22	-10	10
Percent college graduates	0.22	0.15	0.31	-31	104	40

Manchester Lofts

Mean housing value ($)	2,016.25	2,759.32	21,523.26	37	680	967
Mean gross rent ($)	128.00	134.00	268.00	5	100	109
Percent rental units	0.80	0.75	0.57	-5	-24	-28
Percent vacant/other units	0.02	0.08	0.16	321	88	691
Average household income ($)	8,410.83	7,746.63	29,483.33	-8	281	251
Population	1,972	1,455	481	-26	-67	-76
Percent African American	0.89	0.94	0.80	6	-15	-10
Percent college graduates	0.04	0.02	0.14	-48	635	282

FIGURE 9.2. Fulton Gas Works: View of main building along Williamsburg Avenue.

Road, which had been shuttered since the 1950s.[1] The complex had provided gas for heating and cooking to Richmonders in various incarnations since 1854. Over the years, the neighborhood has been partially redeveloped, with new housing, industry, and parks. The one constant has been that gas works, a hulking, abandoned complex of metal and concrete overseeing the destruction and partial rebuilding of a community—all the while vacant itself (see figures 9.2 and 9.3).

As the only city-owned HI-TOAD site in Richmond, city efforts here differed greatly from the other, privately held HI-TOADS. The first official effort to reuse the site was in 2001, when city officials offered it up as a site for the National Slavery Museum. Politics drove the museum to Fredericksburg, Virginia, leaving the gas works behind.

Two major issues have confronted the city as it considers a reuse: the site's location in a floodplain and the unknown level and extent of contamination. The low-lying site is not well suited for redevelopment and a number of prospective buyers have indicated so to city officials. "I've seen it flood up to eight feet in that area," commented one community leader. Rather than explore a passive use for the site, Richmond officials are focused on bringing jobs and property tax-generating use to the site.

Intertwined with city efforts to identify an economically viable use for the site is a fear among city staff about the kind of contamination that may exist on the site. One official remarked, "we have no idea what's out there." Such fears pervade the surrounding Fulton neighborhood and have been a source of worry for some in the area.

City efforts to do something with the gas works continued, and in 2004, the city announced that it would partner with a private developer to create an entirely new community at the site, with both residential and retail uses. Despite the historic qualities of the gas works, the plan involved razing all of the structures and building from the ground up. Eventually, this plan, too, fell through. An article in the *Richmond Times Dispatch* described the site as subject to "lots of ambitious planning but little action so far" (Ress 2004, A-1).

In my interviews with city officials, they continued to be optimistic. One official said that they are "studying, looking at what we can do with it." Another offered some concrete promise: "I'm meeting tomorrow with a national group and [we'll] see what happens, we had a couple folks that got close."

In 2006, the city again proposed a bold plan, to build a minor league baseball stadium on the site. The project quickly stumbled as questions

FIGURE 9.3. Fulton Gas Works: Main entrance to the complex, looking south.

grew about the site's location relative to highways and downtown and the city's inability to answer questions about environmental conditions.

Independent of city actions, the Greater Fulton Civic Association embarked around 2000 on an effort to develop a greenway along Gillies Creek and its confluence with the James River. In their plan, the gas works site is to be used as a museum where visitors can learn about the process of converting coal into gas. Many of the association's ideas were adopted in the 2000–2020 Richmond Master Plan, but all references to the gas works were removed. In figure 9.4, I present a neighborhood map from the Master Plan, with the Fulton Gas Works indicated by an asterisk. The city's plan connects existing parks in the area, from the Great Shiplock Park in the west, to Libbie Hill Park, Chimborazo Park, Gillies Creek Park, and Powhatan Hill Park further east. The gas work's location along that greenway would appear to make it a strong candidate for inclusion in the concept plan, but according to representatives from area community groups the city was not interested.

Fulton Gas Works is at the edge of its census tract, comprised of a mix of older neighborhoods and the Fulton Redevelopment Area alluded to previously. Since 1980, the city has slowly added single-family homes to the neighborhood and the result has been a generally positive direction for the tract relative to the city as a whole. Housing values and rents have both grown at a comparable rate as the entire city, with the mean rents and values slightly higher than for the city. Table 9.2 shows that the population of the tract almost doubled from 729 in 1980 to 1,439 twenty years later (compared with the city as a whole, which lost 10 percent of its population during that same time period). One interpretation of table 9.2 is that those new residents were mostly African American, had more children, had relatively high incomes, and were more educated than the existing residents.[2] These demographic changes were spurred by the city's redevelopment activities in the cleared portions of the Fulton neighborhood and have positively affected housing values and rents. The neighborhood's transformation happened in spite of the purported negative influence of the Fulton Gas Works site.

A half-century after its closure, the Fulton Gas Works stands as a testament to both the challenges of HI-TOADS redevelopment and the city's reactive nature. Real estate developers have driven much of the attention at the site and the city has simply reacted to those proposals. It is evidence of a broader trend in Richmond redevelopment, not just at the gas works.

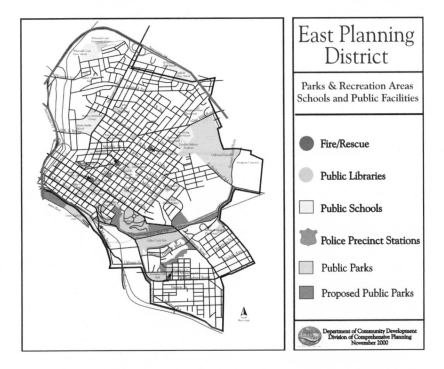

FIGURE 9.4. Detail of 2000–2020 Richmond Master Plan: Fulton Gas Works is located in a central location for open space nework, but the site was not selected to be part of the greenway. 2000–2020 Richmond Master Plan.

"Developers have driven this stuff—that's how Richmond has done things for as long as I've been here," remarked one official. The story of Fulton Gas Works is a sad story of repeated failure. While the city has acted here, its action was not demonstrative of a leadership or proactive or planning style, but of a passive, reactive style. The results of such an approach are clear.

Richmond Memorial Hospital

From 1957 until 1998, the Richmond Memorial Hospital was a key institution in the North Side of Richmond. Situated in one of Richmond's most affluent neighborhoods, the hospital differs from many of the other HI-TOADS examined in this study, which are located primarily in low-income areas. But even in wealthy areas, the abandonment of a 14.5-acre, fifty-year old complex of former health-care buildings can have detrimental effects (see figure 9.5). Local officials and community leaders expressed concerns about illicit occupancy and use of the complex and worried about the uncertainty surrounding its reuse. As with other HI-TOADS, the surrounding neighborhood is also at increased risk of fire associated with the hospital site. While hardly a "poster child" for HI-TOADS, we can learn much from studying reuse planning efforts at a site in such a well-off community.

When the hospital administrators announced the complex's closure in 1998, they promised to work with the surrounding neighbors to create a reuse that was compatible with their needs and to address their concerns. Jeffrey Cribbs, the executive director of the hospital, was quoted as saying "our board has worked a long time to find an *appropriate* use for the hospital" (emphasis added; Rayner 2001a, B-1). To that end, after three years of vacancy, they sold the complex to a consortium of Lutheran churches for $100,000 to develop a retirement facility.[3] The move was supported by neighborhood groups, who wanted a nonprofit organization to take over the property and "believed a retirement community was the best use for the site" (Rayner 2001a, B-1).

Unfortunately, the Lutheran group was incapable of making the reuse work and announced less than a year later that they were abandoning their retirement facility project. The Lutherans felt that the $100,000-per-month maintenance costs were too burdensome and again committed to the local community to transfer the property to a good steward: "We hope to put it in the hands of someone who will do something that is agreeable to the community" (Rayner 2001b).

But maintenance costs were not the only challenge faced by the Lutheran group. They also faced a tough set of neighborhood groups that Carol Holmquist, manager of the development, blamed for the failure: "We did not reach the kind of compromise we needed to with the neighborhood, between what we needed to do and what the neighbors felt they needed" (Rayner 2001c). Part of the Lutheran group's proposal was that some multi-family residential uses be allowed on the site. The neighborhood's lead activist, Norma Murdoch-Kitt, explained that neighbors did not oppose a retirement community, but preferred that most of the structures be demolished and the site be redeveloped for single-family homes (Rayner 2001c). The community's opposition to multi-family housing persisted when the Lutheran group sold the property to a local real estate developer, Shane Parr.

Parr's plan included condominium units and a mix of health-care, educational, and office uses. After purchasing the complex, he was optimistic

FIGURE 9.5. Richmond Memorial Hospital.

about working with the community. "The neighbors are sort of like our partners . . . we want to make them happy" (Rayner 2002). But after several years of difficult negotiations, Parr also failed to advance a reuse of the complicated site. A local business leader explained: "the [neighborhood] association killed the deal, [the developer] wasn't allowed to rent the condo units—still sitting there." Neighborhood opposition centered primarily around the rental of multi-family property, a common NIMBY issue in affluent neighborhoods.[4] In criticizing how much power neighborhood groups have in Richmond redevelopment, one business leader commented that the hospital project was "a perfect example of a neighborhood association: rental is bad, ownership is good." Despite a planning commission endorsement in 2003, the neighborhood groups were able to quash Parr's project because it allowed for multifamily rentals.

A review of census data shows that despite the facility's closure in 1998, there was little impact on 2000 census variables in the host tract. Both mean housing values and mean gross rents grew at a rate comparable to that of the city as a whole, with housing values and average income levels higher in the host tract in 2000 than the city. The neighborhood has a slightly lower percentage of African Americans and a higher percentage of college gradu-ates than the city as a whole.

With the hospital closure coming so close to the decennial census in 2000, it is likely that the externalities generated by the HI-TOAD site had not yet begun to register. In fact, reported housing values in the 2000 cen-sus derive from 1999 figures, only one year after the closure. While the neighborhood today remains vital, rents and housing values may under-perform the city as a whole in 2010, in the event that the site is still vacant.

While the census data points did not indicate so, local officials and com-munity leaders indicated their impressions that the vacant complex contin-ues to exert negative externalities on the surrounding area. Except for the planning commission, local government has been largely absent from this series of private-party transactions. A leading city official asked, "do we need to now interfere? If private sector can't do it themselves, public sector needs top give it a jump start." The city plans to get engaged in the project if the private parties cannot move quickly to advance the reuse. Eight years after the hospital closed, the city now finally does seem poised to intervene.

Tobacco Row

Dubbed the biggest urban-redevelopment project of its kind in the United States, William H. Abeloff's adaptive reuse of fifteen former cigarette facto-ries in the Shockoe Bottom neighborhood of Richmond began as a bold vi-sion. When he began in 1981, the obsolete structures covered 15.5 acres and millions of square feet running between Main Street and Cary Street. By the mid-1990s, Abeloff had already successfully converted four of the western-most buildings into apartments. Over the last decade, all but one remain-ing structure has undergone renovation—the Lucky Strike Factory (see fig-ure 9.6). As a whole, the site continues to meet the HI-TOADS classification, but clearly there is much to celebrate at Tobacco Row.

Mr. Abeloff purchased the buildings for a relatively low price because the immediate neighborhood had a reputation for criminal activity. "Years ago if you kept driving down Cary Street at night it was because you were lost or up to no good. Abandoned warehouses and dark alleys sat as omi-nous reminders that the area from 19th Street to Pear Street, between Main and Dock Streets, had seen its best days a long time ago" (Humes 2003, D-19). A local official recounted to me that prior to the redevelopment efforts,

FIGURE 9.6. Tobacco Row: The Lucky Strike building, looking west.

"all of this had been an eyesore." Another official commented that the beautiful, historic buildings were "chock full of asbestos and creosote."

Built between 1890 and 1940, Tobacco Row offers a glimpse into the urban space that helped Richmond grow and thrive into the metropolis that it is today. The string of massive brick buildings have been largely redeveloped and have the feel of a museum, preserved from a distant past (see figure 9.7). The adaptive reuse of fourteen of the Row's fifteen structures was no accident—it was an essential part of developer Abeloff's plan for the district.

In assembling the financing for the first stage of the Tobacco Row project, Mr. Abeloff encountered difficulties in securing the needed federal historic tax credits. He fought out the battle in the U.S. Capitol, securing the necessary credits and $100 million in tax-exempt bonds for the Tobacco Row project (Drape 1991). To sweeten the deal, the City of Richmond agreed to spending $6.5 million for new streets, landscaping, a new public park, and tax abatements (Drape 1991). The combined public-sector aid was enough to get Mr. Abeloff started.

In the following years, Mr. Abeloff continued to convert more properties, moving east along Cary Street, leveraging the federal incentives, along with growing state and local incentives to redevelop historic properties.

FIGURE 9.7. Tobacco Row: Renovated buildings, new sidewalk paving, trees, and street furniture along Cary Street.

Eventually, Abeloff faced some challenges and sold the remaining vacant buildings on Tobacco Row (including Lucky Strike) to a national real estate developer, Forest City.

Forest City struck a deal with the city that included a $575,000 "incentive loan" and a commitment to build a $1.3 million parking deck at the site (Hickey 1998). Forest City held their end of the deal and proceeded to steadily redevelop the remaining properties heading east along Cary Street. The final, easternmost building is the Lucky Strike Factory, which has sat vacant for more than a decade.

Built to make the famed Luckies cigarettes, the historic Lucky Strike building is last on the Row and, hence, the last to be redeveloped. After taking over the property, Forest City announced that Lucky Strike would house a grocery store. That plan fell through and local business leaders today say that the building eventually will be converted to high-end residential use. While the success down Tobacco Row is significant, the abandoned and uncertain future of the Lucky Strike building poses a continuing challenge to the neighborhood's viability. As a columnist for the *Richmond Times Dispatch* noted about the greater downtown residential area, "both hope and despair always seem to be lurking just around the corner" (Williams 2003, H-3).

As a whole, Richmond performed very well in terms of growth in mean housing value and mean gross rent from 1980 to 2000. Likewise, the census tract including Tobacco Row did well on both indicators, actually surpassing citywide growth rates for both rents and housing values. Absolute housing values in the Tobacco Row area are a fraction of citywide values, $18,720 versus $53,428 citywide, but mean gross rent for the neighborhood exceed the city rents, $600 per month versus $543 per month citywide. These disparities reflect the abundance of rental properties in the renovated Tobacco Row buildings (most of the for-sale condominium were sold after 2000 while Forest City was the primary developer). The census data shows the neighborhoodwide demographic changes underway from 1990 to 2000, in particular: rise in percentage of rental units, rise in total population, rise in household income, decrease in percentage of African Americans, decrease in the percentage of children, and sharp rise in the percentage of residents with college degrees. These trends all run counter to citywide trends and reflect the gentrification at work in the tract.

Interestingly, the incomplete renovation of Tobacco Row may partially explain why the tract continues to dramatically underperform the rest of the city in housing values. When and if the Lucky Strike building is renovated and the uncertainty and dangers of vacant properties are removed from Tobacco Row, then aggregate housing values for the neighborhood may reflect the changes afoot between Main and Carey Street.

How Are HI-TOADS Addressed in Richmond?

Richmond is different in many ways from the other cities included in this book. While Richmond has much in common with Trenton, Pittsburgh, New Bedford, and Youngstown, it is also quite different in terms of percentage of HI-TOADS in city ownership, Also, despite being classified as a "weak-market" city, Richmond is part of a thriving metropolitan area.

Of the six HI-TOADS in Richmond, only Fulton Gas Works is owned by the city. This is in contrast to the other study cities, where most HI-TOADS were received by the city through tax delinquency.[5] With less direct liability, the city of Richmond has less desire to involve itself actively in HI-TOADS reuse, as evidenced in the profiles of Tobacco Row, Fulton Gas Works, and the Richmond Memorial Hospital.

Richmond was identified by the Brookings Institution as a weak-market city, but only one of three such cities located in economically strong metropolitan areas (Vey 2006).[6] This unique quality makes Richmond an attractive location for real estate investors and has played an important role in the city's attitude toward HI-TOADS.

The city's attitude can best be expressed as "reactive." When developers at Tobacco Row approached the city, funding and support were offered. When developers proposed a plan for Richmond Memorial Hospital, the city supported them. When developers looked for the city to partner in re-developing Fulton Gas Works, the city agreed. But, city acquiescence and tactic approval was insufficient at both the hospital and gas works sites. For those sites, and others in Richmond, city leadership and proactive action is needed. Instead, the city has chosen a different path.

Part of the current city approach is a voluntary, business-friendly historic-preservation program. Through a concerted effort to aid property owners in the listing of their structures on the National Register and a series of generous tax incentives, the city has set up a strong incentive to reuse and rehabilitate historic structures. In part due to this program, Tobacco Row, Tredegar Iron Works, and Manchester Lofts were all reused successfully. In addition, the city has maintained a strong code-enforcement program with the express purposes of encouraging adaptive reuse. One city official explained: "Paul Cooper, retired Commissioner of Buildings, enforced section 96 of building code to help reuse historic properties. He was innovative and looked at codes, realized you had to meet them but worked with owners . . . He was open to alternate solutions."

The Commonwealth of Virginia has ambitious programs to address brownfields. Through grants, loans, marketing, and technical assistance, the state aids local governments and private developers with the reuse and redevelopment of brownfields. Oddly, among the state's list of seventy-five official brownfields, only one is located in Richmond.[7] In fact, the state's brownfields office has little direct contact with city officials.

It is the state's Historic Preservation Department that is most involved in redevelopment in Richmond. Through an innovative 1997 historic preservation law, the state provides some of the most generous tax breaks in the country to developers of historic properties. While the federal program supports adaptive reuse for rental properties, the Virginia law allows for homeownership projects. Given the current demand for condominiums in Richmond, this added provision makes a big difference for developers.

The federal government supports HI-TOADS reuse in Richmond primarily in two ways: the U.S. Environmental Protection Agency funds support brownfields assessments and technical assistance, and the Department of Interior's National Register of Historic Places provides a framework for preserving Richmond's historic fabric. The Department of Interior's historic preservation tax credits have provided a direct and measurable incentive to much of Richmond's redevelopment. Both the Tobacco Row and Manchester Lofts projects benefited substantially from the federal tax credits.

In the example of Tredegar Iron Works, the city, the state, and the federal government (through the National Parks Service) worked closely with private property owners to reuse the site effectively. One federal official explained, "it took all the public and private layers together . . . there's power in working together." But such collaboration is rare in HI-TOADS reuse projects in Richmond. Tredegar was the exception and not the rule.

Richmond has a strong and vital network of neighborhood associations and other nongovernmental organizations with scores of groups actively involved in the affairs of local government. The challenge the city faces is how to move forward to plans and projects given the power and influence of these groups.

When he returned to office in 2006, Douglas Wilder instituted a new framework for organizing citizen involvement in city decision-making: the Mayor's Neighborhood Roundtable. Comprised of a citizen representative from each of the nine council districts, this group provides a manageable method for the mayor to hear from his constituents without being overwhelmed by them.

Three active Richmond organizations located in neighborhoods with HI-TOADS were researched: the Greater Fulton Civic Association (Fulton Gas Works), Ginter Park Residents Association (Richmond Memorial Hospital), and Shockoe Bottom Neighborhood Association (Tobacco Row and Richmond Cold Storage). Each took on some role in advocating for reuse of their local HI-TOAD site, but most were reactive in nature—typical NIMBY behavior.

Of the six HI-TOADS included in this case study, Manchester Lofts and Tredegar Iron Works have been reused fully and Tobacco Row is almost fully reused. That leaves Fulton Gas Works, Richmond Cold Storage, and Richmond Memorial Hospital vacant and unutilized. Between my interviews and census data analysis, it appears that the debilitating externalities generated by the reused HI-TOADS may have ended. It is also clear that for those three HI-TOADS that remain vacant, they continue to bring down their surrounding neighborhoods in a variety of ways. City, state, and federal efforts directed at Tobacco Row and Tredegar Iron Works were successful. City, state, and federal historic preservation policies provided the needed incentive for the reuse of Manchester Lofts. But a range of complexities at the remaining three HI-TOADS have amounted to a static situation, where reuse and redevelopment are unlikely in the near future.

While the city deserves some credit for its agreeability to support Tobacco Row, it largely has been reactive in response to developers' interests. Until the city takes on a more aggressive posture to its HI-TOADS, the three remaining sites likely will lay fallow.

Key Themes in Richmond HI-TOADS Redevelopment

Urban policy and planning has always been different in the South. Silver's (1984; Silver and Moeser 1995) and others' work has consistently shown that issues of race and the aftermath of the Civil War, Reconstruction, and the Civil Rights Movement influence redevelopment in southern cities quite differently than in other regions of the country. In closely examining the city's policies toward HI-TOADS over the last decade, there is a remarkably lower level of direct application of policies and planning than in the other case-study cities. I must pause when considering the question about which policies Richmond drew on in addressing its HI-TOADS. The answer is not so obvious.

Throughout this chapter, I have described the various ways that the city has supported (or thwarted) HI-TOADS reuse. But ultimately, the efforts have been minimal in contrast to other cities under study in this book. City efforts at Fulton Gas Works and Tobacco Row were aimed directly toward economic development, as were city efforts at Manchester Lofts and Tredegar Iron Works. But the city has been all but absent from efforts to redevelop Richmond Memorial Hospital and Richmond Cold Storage. Local officials were interested in keeping the structures in some sort of use as a way to promote public health and well-being—not to enhance economic welfare.

In sum, the city's strategies to address these sites were weak and timid. In some cases, the city took no formal action at the sites. As I will explain in the next section, the city's actions over the last decade were directed primarily at changing the basic rules that governed land development to facilitate HI-TOADS reuse—unlike other cities that dove directly into demolition, rehabilitation, reuse, and marketing of sites.

While local government action at individual sites was feeble, Richmond's planning function in general is quite strong and vital. The city consistently produces master plans that are thorough and comprehensive, involve meaningful citizen involvement, are codified by City Council, and are distributed widely. There are two challenges to what otherwise would be meritorious planning: HI-TOADS are excluded from the city's planning and little attention is paid to preserving industrial land within the city.

Written in a typical style of current master plans, the 2000–2020 Richmond Master Plan relies much on generalities and citywide policies and places little emphasis on individual sites or properties. While a few specific sites are discussed in the plan, no HI-TOADS are included.

The Greater Fulton Civic Association's plan to develop a greenway through the Fulton Gas Works site is a good example: the Master Plan includes the greenway but references were removed: "that part didn't make it to the Master Plan, it didn't survive in the plan," explained a community leader.

As a consequence, the city's extensive planning has done little for the most-serious abandoned sites in the city. In an interview with a senior city official, when I mentioned that none of the city's HI-TOADS are included in the Master Plan, he responded "that's why I never look at it." He is not the only one disappointed by the plan. A leading real estate developer said "we go up against the Master Plan every place we turned. The market is changing so fast" the Master Plan can't keep up.

So much planning, yet redevelopment outcomes are mixed. A city official was more blunt: "the Master Plan does not address economic development . . . [and the] need to retain industrial land use." If anything, some city critics contend that the progressive historic-preservation initiatives actually worsen the city's position with respect to preserving industrial land for economic growth. As the logic goes, there is only a finite amount of acreage within the city borders for which industrial activities are allowed. Most of that acreage is improved with historically significant buildings, which, for the most part, do not meet modern facility requirements for industry. By discouraging demolition of these structures, the city, state, and federal governments are providing an incentive for nonindustrial use of buildings in currently industrial zoned areas.

While planning should consider this, Richmond has done little to face this paradox. The Manchester neighborhood, where the Manchester Lofts conversion took place, is one of the last remaining industrial hubs in the city. As businesses leave outdated and obsolete structures there, residential developers are keen on moving in and adaptively reusing them for apartments and condominiums. The long-term effect is a shrinking of total industrially zoned land in the city. That may be appropriate, given the wider trends at work in the region, but considering such effects should be a big part of planning and it is lacking in Richmond.

. . .

Some American cities are aggressive when it comes to derelict and abandoned properties. Other cities are more laissez-faire, letting the market dictate land-use outcomes. Richmond is in between these two extremes. The framework that organizes Richmond's approach to HI-TOADS is based on a citywide proactive strategy to encourage the adaptive reuse of historic structures, while taking a reactive posture to the development of individual sites.

Closely mirroring the federal and state historic-preservation programs, the city has in place a generous suite of tax breaks to encourage adaptive reuse. The city also uses an aggressive code-enforcement program to encourage historic preservation. The two strategies do not identify any particular properties or districts for prioritization. In an effort to appear fair, the programs also fail to acknowledge the varying levels of blight that different types of historic properties generate. That is, an occupied historic mansion that is in need of renovations qualifies for the same public support that an abandoned industrial plant might. However, the neighborhoodwide externalities generated by an abandoned industrial plant might be more severe than that of an occupied mansion in need of a facelift. Of course, measuring that distinction can be difficult and the city's approach has rested largely on a citywide approach irrespective of neighborhood impacts.

By setting the table for adaptive reuse, city officials have deliberately avoided engaging directly with promoting the reuse of individual sites. This approach is similar to what happened in New Bedford, where a brownfields task force listed the top brownfields sites in the city and then retracted the list when property owners complained.

The other reason for avoiding individual sites is the city's fear of liability. "They, like most cities, play their cards close to their chest. They think that if DEQ [Virginia Department of Environmental Quality] finds out about one of their sites they will get slammed" commented a state official.[8] By keeping their plans to themselves at a site like the Fulton Gas Works (for over fifty years), the city has not exactly advanced the reuse of that site. Cities need to be open and active about their plans, commitments, and strategies about reuse of HI-TOADS in order to effectively engage property owners, other stakeholders, and the public in a decision-making process.

But such active government involvement does not always fly in a conservative city like Richmond, where property rights are strong and government meddling in individual sites is often considered anathema. This makes the reuse of HI-TOADS challenging, but not impossible. The private-developer-initiated projects at Manchester Lofts, Tredegar Iron Works, Richmond Memorial Hospital, and Tobacco Row speak to the potential of the private sector in addressing HI-TOADS. And where the city feels that market mechanism are not working well enough, there is clear interest by the city to intervene as a last resort. There is much that other cities can learn from Richmond's example.

Richmond's laissez-faire policies have roots as far back as Thomas Jefferson and his land-based views of democracy that shaped the U.S. Constitution. For centuries, Richmond was the anti-urban capital of the South, where government left matters of building and land use to private property owners, while maintaining a profound respect for historic preservation (Tyler-McGraw 1994). In fact, the historic preservation movement itself was born in Virginia in 1853, when the Mount Vernon Ladies' Association mounted an effort to preserve George Washington's residence at Mount Vernon, Virginia (Stipe 2003). But the state's commitment to historic preservation and property rights began to change in the mid-twentieth century with the advent of urban renewal.

The widespread seizure of private property, bulldozing of entire neighborhoods, and destruction of much of the historic fabric of the city during the urban-renewal era left many Richmonders and Virginians wondering what had happened (Tyler-McGraw 1994). In an interview with a senior city official, he argued that the city and state historic-preservation policies were "a reaction to the kind of demolition that happened in the state's urban centers in the 1950s and 1960s—it was a grassroots effort." Fed up with urban renewal and better connected to its roots as a center for historic preservation, both Richmond and the Commonwealth of Virginia embarked on a major change of course with respect to redevelopment.

This new approach was grounded in a nonregulatory, industry-friendly approach to historic preservation. By establishing a framework to provide financial support to real estate developers, the state advanced a number of public policy objectives simultaneously: reuse of abandoned buildings, increased economic activity, and preservation of historic resources.

In 1995, the Virginia Department of Historic Resources tasked its director of historic preservation, Kathleen Kilpatrick, to develop a state equivalent of the federal historic-preservation tax–incentive program. Working closely with statewide preservation groups and other state agencies, she made a pitch to then-governor Jim Allen, a free-market-oriented Republican. According to my interviews, Governor Allen liked the idea, but was reticent about making it his own initiative, given the competing demands. He was observed by a state officials to have said about the bill, "you get it introduced and I will support it."

With that, Ms. Kilpatrick galvanized much interest among legislators to develop a program that would be voluntary, but generous. It was viewed widely as a way to increase services to the state in support of preservation, while doing so in a way that was consistent with the state's strong property-rights bent. Ultimately, the law quietly passed and was signed into law by Governor Allen. In describing the unique qualities of the new law, one state official said: "Virginia is a state unlike Connecticut or Rhode Island that has a deep traditions of really caring about its resources and really being suspicious of government. Our programs rely on incentives, rewards, recognition, but are non-regulatory—limited government scenario."

These incentives were designed to help redevelop the core of Richmond. A city government leader said, "the tax credits are the financial tool to make it happen. The tool to create historic districts." The law also addressed some financial barriers that real estate developers had faced in the past working with the federal program. In particular, the Virginia program provides a greater degree of flexibility in assigning the credits and converting them into cash.

Ultimately, the new law has been remarkably successful in encouraging the reuse of historic properties, HI-TOADS and non-HI-TOADS alike. One business leader explained that with respect to redevelopment, "what's happened in Virginia, in specific in Richmond, [is a result of] a combination of federal tax credits, the Virginia state tax credit and redevelopment tax abatement program [the city's program]." A state official concurred: "I think that the state rehabs and federal tax credits are the reasons that these projects have come to fruition."

Private real estate developers are often at odds with community-based organizations (CBOs) in urban redevelopment. The case of Richmond is no different. CBOs often conceive of reuse strategies, get local government involved, and recruit developers. This is where Richmond differs. The home of many active CBOs, Richmond has a strong network of neighborhood-based groups. These groups had little to do with HI-TOADS and if anything, they are reactive to new development projects, like the city government. Instead, it is private real estate developers who drive the action in Richmond.

The typical CBO response to HI-TOADS is to sit back and react to others' efforts. A CBO leader said to me, "we want to see these properties fixed up—our neighborhood has been challenged." He continued: "It's really a no-brainer, it's just going to help the community. The more people who live here, the more attention we'll get from the city. Now, 'crackheads' break in and live there [in Richmond Cold Storage]. Those aren't the kind of people I want living in my neighborhood." But, ultimately, his organization has not actively attempted to advance the redevelopment of neighborhood HI-TOADS. In the more-affluent neighborhood hosting the Richmond Memorial Hospital, the neighborhood groups actually have been blamed for stymieing the reuse of the site by demanding too many concessions from developers. The other CBOs in HI-TOADS neighborhoods fall somewhere in between those two extremes, either neutral to reuse or an impediment.

Real estate developers have responded to this variation by avoiding redevelopment in certain areas. One business leader told me, "I've tended to go for neighborhoods where there is no association or they're not strong. If associations are against a project in Richmond, the City Council votes it down." As a result, developers avoid places like Fulton (where Fulton Gas Works sits) and Ginter Park (home of Richmond Memorial Hospital). And developers have flocked to Manchester and Shockoe Bottom due to their relatively weaker CBO presence—the Tobacco Row neighborhood had been entirely industrial prior to the redevelopment and had no immediate residential population. The consequences are clear: Developers are the drivers behind action at HI-TOADS in Richmond and CBOs are neutral at best, impediments at worst to reuse.

Motivated largely by an economic development policy framework informed by an interest in historic preservation, the City of Richmond has had some success in reusing HI-TOADS. As a city with long ties to historic preservation, the adaptive reuse of historic structures makes much economic sense. Ironically, the city's biggest critics point to the shrinking industrial acreage in the city as the central flaw in historic preservation. From the perspective of HI-TOADS reuse, historic preservation in Richmond works well to get some derelict and dangerous sites into productive reuse. The long-term impacts on Richmond's local economy are less clear.

The city and state's strong historic-preservation programs appear to have been a backlash to the tragic demolitions of urban renewal. Public programs have laid a solid foundation for private developers to reuse and rehabiliate buildings and neighborhoods and in a way that increases tax renevues for the city, brings in new employment, and increases the residential population. Not only that, but these adaptive reuse projects have tended to bring a new kind of residential population back into Richmond: wealthy, well-educated, and primarily white gentrifiers.

There are few examples of the persistent racism and remnants of the "separate city" in HI-TOADS reuse. Redevelopment in the Tobacco Row and Manchester neighborhood speak to a new racial calculus in Richmond where anyone is being welcomed back into the city (especially those with money) in order to stem the city's continuing population decline. A different story emerged in Ginter Park, a majority-white neighborhood, where fear of new multi-family housing intimates a fear of the "other."

With African-American leadership firmly in place within the city government, the legacy of past discrimination and racism are manifested through contentious citizen participation and community-based organization involvement. The scope of CBO involvement, the power of these organizations, and their relative disinterest in playing an active role in reuse, are also relics of the top-down, autocratic urban-renewal programs. Now, that top-down process is replaced by a developer-driven, city, and CBO-reactive redevelopment process that has results in much HI-TOADS reuse.

[10]

Conclusion

Every day, HI-TOADS present a multi-faceted, seemingly intractable challenge to local officials and community leaders trying to revitalize distressed urban neighborhoods. Allowing these sites to fester imposes unfair burdens on surrounding residents and inhibits neighborhood well-being. Doing something about them can stabilize neighborhoods, planting the seeds for improved quality of life. This book has been about the stories of community action and the results. Here, in this final chapter, I return to the original issues that spurred this book and review the potential this work has for communities facing these same challenges. Like any scholarly work, this book has limitations and I report on those and how future research could address them.

Beginning with the premise that HI-TOADS are unwanted, polluted, and dangerous, this book is about what can be done about them. To answer that question, this study first identified those cities most likely to have multiple neighborhoods with HI-TOADS. The ranked list appears in Appendix B. The cities at the top of the list are largely from the traditional manufacturing belt and portions of the old militarized South, cities well known for their decline and a legacy of abandoned heavy industry.

In considering what can be done about HI-TOADS, this study first asked whether local planners recognize HI-TOADS as a problem in the neighborhoods of their cities. The answer is "yes." In the telephone interviews, 35 officials of 38 could identify at least one HI-TOAD site in their city. Local officials were also concerned about the ways in which these HI-TOADS threaten neighborhood stability, cause a wide range of impacts including crime, arson, dumping, and lead to brownlining.

I chose Trenton, Pittsburgh, New Bedford, Youngstown, and Richmond as case-study cities, in part because local officials in those cities demonstrated an immediate understanding of the HI-TOADS problem during the Stage Two phone interviews. So it should not be surprising that the local officials in the case-studies cities showed a high level of understanding about the severity of HI-TOADS.

In chapter 2, I introduced neighborhood life-cycle theory and alternative neighborhood-change theory. Under neighborhood life-cycle theory, local officials view their intervention as necessary to arrest a dying process, analogous to the death of a living being. City officials regularly took action in HI-TOADS neighborhoods using the neighborhood life-cycle theory framework. In only a single case did officials draw on the alternative neighborhood-change theory, in discussing the asset-based planning strategy in Dayton, Ohio. When faced with massive population loss and tragic disintegration of the city's social fabric, city officials have shifted their focus away from the intervention-style strategies that emerge from neighborhood life-cycle theory. Adopting an alternative neighborhood-change-based strategy of asset-based planning, Dayton is able to escape the meta-narratives of urban decline and death and orient their public efforts toward enhancement of community assets.

In the case studies, a demarcation appears between the economic development policies (strongly grounded in neighborhood life-cycle theory) and the community empowerment and healthy environment policies (grounded more in the alternative neighborhood-change theories). Using economic development approaches, officials in each case-study city borrowed from the neighborhood life-cycle discourse on sick or dying cities as justification for intervention—as I expected they would. As a prerequisite to most condemnation actions, cities are required to declare neighborhoods blighted. Blight, a tree disease, is the embodiment of neighborhood life-cycle theory—the notion that changes occurring in a neighborhood are analogous to a disease that afflicts flora.

As expected, community empowerment and healthy environment approaches to HI-TOADS reflected the language of alternative neighborhood-change theories. In these strategies, local people are enabled to address the problems they see in their neighborhoods. The people are the focus, whether in terms of public health or the quality of their living environment. This research contributes to these alternative neighborhood-change theories by showing evidence of where they are put into practice and examples of where such practice is effective. The stories of Magic Marker, Morse Cutting

Tools, and Pierce Mills are particularly powerful in demonstrating the value of grounding planning and public policy in these alternative neighborhood-change theories, rather than neighborhood life-cycle theory.

Of those cities that acknowledge HI-TOADS as a problem, this study asked what policies those cities use to address them. The officials involved in the telephone interviews were fairly uniform in saying that they used economic development and site-work policy tools to address HI-TOADS. Roughly half said they used zoning, planning, and condemnation policies.

In the case studies, I was able to better understand this set of policy tools and their application. First of all, cities do not apply planning and public policies uniformly to all HI-TOADS. While Pittsburgh prioritized HI-TOAD efforts based on a mix of economic development and community empowerment priorities, Trenton's efforts were motivated more by a larger land-use planning framework. New Bedford's efforts were driven by community concerns, Youngstown desperately tried to convert HI-TOADS into jobs and tax revenues, and Richmond left redevelopment mainly to private developers. While cities use different planning and public policy approaches, they apply such strategies where it suits their political needs and constituencies. Most interesting was the extent to which some cities use community empowerment policies to support NGO goals and objectives at HI-TOADS.

Of the three sets of policies, economic development, community empowerment, and healthy environment, the case studies showed economic development (sixteen times) and community empowerment (ten times) were the most frequently used approaches at HI-TOADS, followed by healthy environment (eight times) (see figure 10.1). Despite the rhetoric about economic development, many cities hosting HI-TOADS have few resources for creating new jobs and increasing tax revenues, so they pursue other approaches whenever possible. When approaching a HI-TOAD site, most cities use the same set of economic development tools that they routinely use to attract and retain businesses. But most of the time, the incentives are available only when a company approaches the city. For many of the cities in this study, that means that their economic development strategy involves a great deal of waiting.

When cities have resources for brownfields redevelopment and a willing developer, they focus their efforts on those sites with high market value. In New Bedford, despite all of the external acclaim, the city invested very little

FIGURE 10.1. Case-study sites: Policy applications 1995 to 2005.

	Reused HI-TOADS			Active HI-TOADS		
	Economic development	*Community empowerment*	*Healthy empowerment*	*Economic development*	*Community empowerment*	*Healthy empowerment*
Trenton						
Magic Marker Site		▲	▲	—		
Crane Site	▲			—		
Roebling Complex				—	▲	△
Assunpink Greenway				—		▲
Pittsburgh						
South Side Works	▲	△		—		
Nine Mile Run	△	▲		—		
Hazelwood LTV Site				—	▲	
Heppenstall Site				—	▲	
Sears Site	▲	△		—		
New Bedford						
Elco Dress Factory				—	▲	△
Fairhaven Mills				—	▲	
Morse Cutting Tools		△	▲	—		
Pierce Mills		▲	△	—		
Youngstown						
Aeroquip Site				—	△	▲
YST/RS				—	▲	
Ohio Works	▲			—		
229 East Front Street	▲			—		
YBM Site				—	△	▲
Richmond						
Richmond Memorial Hospital				—	△	
Richmond Cold Storage				—		△
Tobacco Row †	▲			—		
Fulton Gas Works				—	▲	
Tredegar Iron Works	▲			—		
Manchester Lofts Building	△			—		

Code: ▲ Strong reliance △ Some use

† The entire Tobacco Row complex has not been reused; the Lucky Strike site remains vacant.

in promoting redevelopment of its HI-TOADS. Rather, using a community empowerment approach, the city looked for ways to empower and support NGOs in their efforts to address HI-TOADS. In the same way that community empowerment was illustrated in the work of Dewar and Deitrick (2004), the example of New Bedford was that local residents were instrumental in driving reuse efforts and were the true beneficiaries of those efforts.

Trenton's efforts at the Assunpink Greenway and Magic Marker, as well as New Bedford's work at Morse Cutting Tools, Elco Dress Factory, and Pierce Mills, were examples of the application of healthy environment approaches to the problem of HI-TOADS. While almost every HI-TOAD site generates public health and environmental hazards, these were the only five examples where addressing those hazards was a major motivator behind city action. In the cases of the other HI-TOADS, the goals of intervention were grounded in economic development or community empowerment and any public health or environment improvement were incidental.

Part of exploring cities' actions in the study was attempting to understand how HI-TOADS were incorporated into cities' planning efforts. Some cities address planning concerns associated with HI-TOADS through asset-based planning, neighborhood planning, and the establishment of brownfields offices. However, in the Pittsburgh and New Bedford case studies, local officials (with the exception of planners) generally felt that they would lose control to the public if they engaged in planning. Rather, they sought ways to think long-term and comprehensively, and to develop future visions without engaging in a formal public planning process. Trenton was different. There, a formal land-use plan process was developed, although it lacked a truly comprehensive component (a full master plan update is currently being developed).

Returning to the definitions in chapter 2 of planning, planners, and government planning, local officials are engaged in some planning at HI-TOADS, but formal government planning at these sites is rare. While chapter 2 defines planners broadly, there appears to be an important dichotomy between professionals who self-identify as planners (and often have professional planning education) and those who work as planners but do not self-identify as such. The self-identified planners typically work in planning offices and have little involvement in reusing HI-TOADS, while those who do not so self-identify (and often do not have training in planning) are the ones working on the HI-TOADS projects. Therefore, planning, as an activity, is an important component of what cities do to address HI-TOADS, but the planners doing that work often do not self-identify as planners and lack formal planning education.

It is one thing to understand what it is that cities do. It is quite another to understand why. While I did not explore this question in the telephone interviews, I did discover much in the case studies. In Trenton, the cities' actions were driven largely by a commitment by officials to a land-use plan, a vision of what the city will become in the future. In Pittsburgh, the city's place as a center for the region provides a quite different setting than Trenton. Whereas all new developments are subsidized heavily in Trenton, Pittsburgh has a healthy real estate market that makes HI-TOADS ripe for development. For city officials in Pittsburgh, the goal was to maximize property-tax and employment benefits from new development, which meant that the city concentrated its efforts on HI-TOADS with high development potential.

In contrast to both Trenton and Pittsburgh, New Bedford's HI-TOADS policies were driven by a city strategy to become a brownfields capital. Where HI-TOADS were an issue for all five case-study cities, in New Bedford, they were viewed by city officials as a central piece in a strategy to re-invent the city and position it as a leader in redevelopment. City officials saw the limelight as essential to remaking New Bedford.

In the telephone interviews, I found an interesting split between city officials who advocated shovel-ready policies for tearing down HI-TOADS to make sites available for prospective developers and those who saw aesthetic and cultural value in preserving the structures of historic HI-TOADS. The dichotomy was exacerbated in Trenton, Pittsburgh, and New Bedford, cities that all pride themselves on their history and use it to promote tourism.[1] While celebrating the past in museums and tourism brochures, these cities routinely demolished historic structures. Demolition was seen by many as the way to save these cities. New development is viewed by many officials as the way to return these depressed cities to greatness again. Further, shovel-ready sites are believed to be more attractive to developers than improved, derelict sites (no matter how historic they may be). However, this research did not include interviews with developers to validate that conclusion.

While the third question asks about local government action, NGOs' involvement in HI-TOADS is also worthy of study. The nonprofit, Isles, Inc., played a pivotal role in the Magic Marker HI-TOAD site due to a pilot funding project through the State of New Jersey's Department of Environmental Protection. But at Trenton's other seven HI-TOADS, there is a conspicuous absence of nongovernmental interest and involvement. Likewise, in Richmond, with an extensive network of NGOs, I expected that they would be more involved than they were. Youngstown differs from the other cities in that it does not have a viable NGO sector and of the few active organizations, none are involved in HI-TOADS reuse.

The story in New Bedford is quite the opposite, with community groups gathered at almost all of the city's HI-TOADS fighting for demolition, remediation, and reuse resources. In fact, in New Bedford the major actors at HI-TOADS are community groups, not the city. Pittsburgh falls somewhere in between in that NGOs had a catalytic effect at each HI-TOAD site, initiating a project, leading a visioning exercise, or demanding city action. The city government has been the major actor, though having to share the stage with NGOs. Through the city's URA, the city led the reuse of South Side Works, Nine Mile Run, and the Sears Site, while the consortium of foundations and the Regional Industrial Development Corporation of Southwestern Pennsylvania have led redevelopment activities at the city's other two HI-TOADS, the Heppenstall Site and the Hazelwood LTV Site.

The final question of this study was: How successful are city policies at addressing HI-TOADS? The clearest way to measure success when considering HI-TOADS is to ascertain whether community leaders and local officials perceive that a site continues to be a negative drag on neighborhood property values. Active protection and maintenance of a property can be sufficient to mitigate against the externalities generated by a HI-TOAD site. This protection and maintenance function was displayed by property owners in various ways (including fencing, security systems, and signage) in the five case studies. But the best evidence for sound protection and maintenance of a site was the demolition of derelict structures. When funding was unavailable for full site remediation, reuse, and redevelopment of a HI-TOAD site, these protection and maintenance measures went quite far.

In the telephone interviews, I asked local officials about the extent of their success in addressing HI-TOADS. Twenty-four of 35 officials (68.6 percent) said that their city has been successful in addressing HI-TOADS. I further probed this question in the case studies and found the answer to be more ambiguous. In figure 10.2, I present summary census data for all twenty-four HI-TOADS examined in the five case studies. The up arrows

FIGURE 10.2. Case-study sites: Neighborhood performance 1990 to 2000.

	Reused HI-TOADS		Active HI-TOADS	
	Rents	*Housing Values*	*Rents*	*Housing Values*
Trenton				
Magic Marker Site	●	▲	—	
Crane Site	▼	▼	—	
Roebling Complex	—		▼	●
Pittsburgh				
South Side Works	▲	▼	—	
Nine Mile Run	●	▼	—	
Hazelwood LTV Site	—		▼	●
Heppenstall Site	—		▲	▼
Sears Site	▼	▼	—	
New Bedford				
Aerovox Site	—		▲	●
Elco Dress Factory	—		▲	●
Fairhaven Mills	—		▲	▼
Morse Cutting Tools	▲	●	—	
Pierce Mills	▲	▼	—	
Youngstown				
Aeroquip Site	—		▲	▼
YST/RS	—		●	●
Ohio Works	▲	▲	—	
229 East Front Street	n/a	n/a		
YBM Site	—		●	●
Richmond				
Richmond Memorial Hospital	—		▼	●
Richmond Cold Storage	—		▲	▼
Tobacco Row †	▲	▲	—	
Fulton Gas Works	—		●	▲
Tredegar Iron Works	n/a	n/a	—	
Manchester Lofts Building	▲	▲	—	

Code: ▲ Tract outperformed city by at least 10%
  ● Tract did not outperform city
  ▼ Tract underperformed city by at least 10%

† The entire Tobacco Row complex has not been reused; the Lucky Strike site remains vacant. Because the Tobacco Row redevelopment began in the mid-1980s, I present here the census results from 1980 to 2000.

indicate growth in property values and rents at HI-TOADS from 1990 to 2000 relative to each city as a whole and the down arrows indicate relative decline in values and rents. While I would expect to a see a pattern of up arrows on the left side (reused HI-TOADS) and down arrows on the right side (active HI-TOADS), the empirical data from the case studies shows a different pattern. With the exception of Richmond, with its booming economic conditions, sixteen of eighteen reused and active HI-TOADS (89 percent) experienced either lower relative growth in housing values or similar growth levels to their city. Overall, the story is the inverse in examining growth in rents: thirteen of the seventeen (76 percent) sites experienced higher or similar rent growth relative to their city as a whole.

Due to the weaknesses in using housing value data from the census and the coarseness of the unit of analysis (census tracts), I need to rely more heavily on the interviews and direct observations to make sense of the successes of reusing HI-TOADS. In that respect, a number of city actions were particularly effective in addressing HI-TOADS. In the case of Pittsburgh, the city's direct involvement as developer was essential to the reuse of Nine Mile Run, South Side Works, and the Sears Site. In Trenton, the city's funding of demolition and remediation was the most effective means to handle the Magic Marker site. In New Bedford, the city's efforts to bring in external funding and resources were necessary to the demolition and remediation work at Morse Cutting Tools and Pierce Mills. Youngstown's aggressive public acquisition and redevelopment programs at Ohio Works and 229 East Front Street worked to get two critical HI-TOADS back into productive reuse. Lastly, Richmond laid a foundation of historic-preservation programs to provide major incentives for developers to adaptively reuse former industrial properties at Manchester Lofts, Tredegar Iron Works, and Tobacco Row. In sum, there appears to be a strong association between city action and neighborhood conditions around HI-TOADS, but limitations of the study make causality more elusive.

Implications for Planning and Public Policy

These findings have important implications for planning and public policy in five particular ways: (1) the roles that planners and the planning function play in redevelopment; (2) the benefits of community empowerment; (3) balancing historic preservation and demolition needs; (4) local government protection and maintenance; and (5) the prioritization of city efforts. For each implication, I offer policy recommendations. I will describe the political conditions under which cities could consider the adoption of these recommended policies.[2]

The Roles of Planners and the Planning Function in Redevelopment

The activities of public redevelopment vary among different local governments. Many cities have an economic development department, which usually leads redevelopment efforts at individual sites (Zelinka and Gates 2005). Economic development departments work at the scale of individual sites, neighborhoods, and entire cities (Blakely 2000). They work to provide incentives for relocation, tax breaks, and funds for demolition and remediation.

Most cities also have a planning department, which supports redevelopment efforts through special use districts, rezoning, master planning, and other regulatory tools. Planning departments also review redevelopment projects for compatibility with neighboring uses, adequate infrastructure, and in some cases, aesthetics (Meck, Wack, and Zimet 2000).

While planning departments primarily work at the neighborhood or city scale, their development review activity is at the site scale. The HI-TOADS phenomenon by definition is both a site and a neighborhood problem. But it is the economic development departments that most often manage HI-TOADS reuse projects. In many cities that I studied, the power to acquire land and to improve it rested solely in the economic development departments. Cities have vested these departments with the power to do real estate work and essentially have cut the planning departments out of the deals.

Nothing is inherently wrong with economic development departments leading city efforts, but the problem I discovered was that (as their department name implies) economic development departments are primarily concerned with generating tax ratables and new employment. Given that, they are most concerned with addressing HI-TOADS with potentially high market value, rather than the sites that may have the greatest negative impact on residents. While planners work in economic development departments, doing planning, in many cases they do not self-identify as planners and lack planning education.

Planning education teaches professionals how to work with citizens to envision new uses for obsolete land uses. Planning departments have statutory and regulatory tools to move from such a vision or plan to a reused HI-TOAD site. Yet, throughout this research, I discovered that it was often the economic development departments that were in charge of HI-TOADS. In Pittsburgh, a planner complained that economic development officials kept his office out of the early planning for a HI-TOAD site and city planners were directed to rubber stamp the final deal.

Even though this study focused on HI-TOADS, the findings have important implications for understanding the role of planning in urban redevelopment in general. Cities' economic development departments are taking the lead on key redevelopment projects and planning departments are playing a minor role, if any. Why this occurs is not entirely clear. The reason may have to do with the perception that planning departments are not effective at the kind of work necessary to address HI-TOADS. Planning departments were clearly competent at some activities, like writing zoning codes, reviewing development proposals, producing annual reports, and advising the planning board. However, activities in which planning departments' competence was questioned involved more politically charged tasks such as agenda setting, project prioritization, and community visioning. For those tasks, economic development or brownfields offices took the lead.

The exclusion of planning from much of the HI-TOADS redevelopment efforts may be motivated by the desire by local officials to keep redevelopment deal-making opaque. The recent track record of citizen participation in the work of planning departments has made them an unusually transparent component of city governments today (Brody, Godschalk, and Burby 2003). By diverting redevelopment activities away from planning departments, cities are compromising their ability to think, act, and plan comprehensively and in the public interest.

From this research, I recommend that governments at all levels reexamine the role of public participation in economic development departments and implement mandatory planning requirements on these departments. In addition, I recommend that local governments look for ways either to direct economic development departments to consider non-economic measures of their successes (such as improvement of neighborhood quality due to the conversion of a HI-TOAD site) or to empower planning departments to take on redevelopment of HI-TOADS that have deleterious neighborhood impacts but low prospects for job creation or new property tax revenues.

One reason cited by officials for cities' strong orientation toward job creation and property tax revenues in HI-TOADS reuse was the requirements of external funders like state and federal agencies. I recommend that such funders re-examine their strict requirements and provide greater leeway to local governments to address HI-TOADS in ways that may not lead directly to new jobs or new taxes. For example, the reuse of HI-TOADS for passive or active open space, for community gardens, or for a community meeting space would not rank well in state and federal funding applications.

Planning departments generally are staffed with professionals who self-identify as planners and have planning education. These professionals can contribute greatly to HI-TOADS reuse efforts led by other city agencies. I recommend that planning departments market their services to other offices, rather than fighting turf battles. Planning departments should work harder to demonstrate their utility to improve their cities' planning activities.

In more than one example, HI-TOADS in this study could be successfully redeveloped only with substantial infusion of public subsidy. When public resources are low, these sites lay fallow. But in some cases, the assemblage of a financially upside-down HI-TOAD site with abutting parcels could turn around the developer's pro forma analysis and make the redevelopment financially feasible. In such a case, the public sector's role is not to offer developers grants, loans, and tax abatements, but rather to assist the developer in acquiring abutting parcels. When such acquisition is resisted, a city's powers of eminent domain are invaluable.[3]

Where neighborhood conditions are poor due to the blighting effect of a HI-TOAD site, little opposition would be expected to a city's use of condemnation. In cases where HI-TOADS reuse would otherwise not occur or require extensive public subsidy, I recommend that cities actively utilize their condemnation powers to support such assemblage projects.

Community Empowerment as a Policy Choice

Despite attention in recent years to the potential of community empowerment as a tool for local governments to assist and empower distressed neighborhoods, little empirical evidence has been gathered demonstrating its importance (see Dewar and Deitrick 2004). This study showed that in New Bedford, particularly, a community empowerment set of policies were used effectively to address several HI-TOADS.

In New Bedford and Pittsburgh, the NGO sector is vital and was instrumental in both cities' successes with HI-TOADS. In Trenton, community empowerment was only used at the Magic Marker. The impetus for community empowerment at Magic Marker was a State of New Jersey initiative, not a city policy choice. With the exception of Isles, Inc., little capacity exists among Trenton's NGOs for dealing with brownfields and urban redevelopment. The implications of this research are that community empowerment policies can be valuable to support redevelopment efforts, particularly in cases where there is a high level of capacity among NGOs.

These findings contribute to our understanding of how urban policymaking and planning happens. These findings offer further empirical evidence to a refinement of urban regime theory. While an urban regime was fairly unitary in Trenton, the fractious nature of Pittsburgh's development scene suggests multiple regimes in competition—each seeking an inventory of HI-TOADS to redevelop itself. Sites' (1997) observation that community groups working outside urban regimes play an important role in cities is confirmed by this study. Such groups were of particular importance for HI-TOADS reuse in New Bedford.

I recommend that federal and state funding for community empowerment efforts in redevelopment be enhanced. I further recommend that local governments look for ways to channel resources and authority for redevelopment of HI-TOADS to NGOs. The State of New Jersey's pilot project at Magic Marker was strong evidence that NGOs can relieve local governments of some of the burden of managing redevelopment and can help ensure a role for the public in that process.

Given the grounding of community empowerment in alternative neighborhood-change theory, I also recommend that urban planning educators teach such theories to future practitioners. While I studied many successful reuse projects grounded in neighborhood life-cycle theory, it would behoove planning educators to present both theories to students.

Early Warning Systems: Historic Preservation and Demolition

In this research, I advanced our understanding of the challenge of historic preservation of HI-TOADS. The implications of these findings are that HI-TOADS can be reused adaptively and historically preserved if an entity takes action to preserve them prior to or at the time of abandonment. I recommend that local governments consider the adoption of an Early Warning System (EWS).

EWSs have been used by housing planners to predict housing abandonment (Hillier et al. 2003). A HI-TOADS EWS would be a simple tracking system where large commercial and industrial sites with a high probability of abandonment are monitored. City officials quickly learn through word-of-mouth when a site is on the verge of abandonment and when it is abandoned. If this information is consolidated in a tracking system, then city property managers could be deployed to an abandoned site immediately. When owners are still present, city officials can use an EWS to engage them and cooperate on a property management plan while the site is unoccupied. In the case of absentee or missing owners, the EWS can aid the city in initiating expedited measures to take some level of control of the property for the purposes of protection and maintenance. Basic protection and maintenance of historic properties can help preserve them for possible future adaptive reuse.

Local Government Protection and Maintenance

While local government–administered protection and maintenance of historic HI-TOADS is essential, even nonhistoric HI-TOADS need protection and maintenance to mitigate their neighborhood impacts. However, local government–led protection and maintenance of HI-TOADS is not common. This has implications for the administration of cities. A new role appears necessary for cities to take on as property managers for properties not owned by the city. While legal, political, and financial barriers have prevented cities from taking on such a role, these findings suggest that tackling those barriers is essential for cities to address HI-TOADS effectively. I recommend that cities commit to protecting and maintaining HI-TOADS through an active program of security, fencing, demolition of derelict structures, and trash removal. In cases of known or suspected environmental contamination, such activities would need to be curtailed to reduce harmful exposure to city workers.

Since many HI-TOADS end up in local government ownership as a result of tax foreclosure, cities should view HI-TOADS as financial assets before they possess them. Early and continuing investment of protection and maintenance funds in HI-TOADS make these sites more appealing for reuse or redevelopment, while also ameliorating the more severe neighborhood impacts at the sites (like dumping, squatting, or criminal activity). Of course, it is dangerous for cities to take on such liabilities prematurely, especially when environmental hazards are unquantified.

Many cities, such as Cleveland, Ohio, have established land banks designed to dispose strategically of tax-foreclosed residential property (Dewar 2006). Land banks could be a good avenue for cities to develop, finance, and execute property management plans for HI-TOADS, with funds from the sale of HI-TOADS going back into the land bank to cover the costs of future protection and maintenance. New enabling legislation at the state level would be needed in most states to extend this land-bank authority.

Another model for cities to consider is the federal Office of Real Property Asset Management, within the U.S. General Services Administration (GSA). The federal Property Act (40 U.S.C. 543, 116 Stat 1062) requires that the GSA protect and maintain all excess and surplus government-owned real property until it is disposed of. The law also provides GSA a chance to lease out the property in order to generate revenue while it is in the process of disposing of the property. The proceeds from the sale of excess and surplus property in some cases can be retained by GSA for the purposes of running the disposal program (including protection and maintenance) or for new development projects (Matthews 2006; Board on Infrastructure and the Constructed Environment 2004). Local governments should explore the possibility of developing comparable laws and programs within their cities for managing the reuse of HI-TOADS.

Hazard Ranking System

HI-TOADS are prioritized for redevelopment based upon a variety of political, economic, and social reasons. What is clear from the research is that the way that cities consider those reasons is entirely subjective and lacks transparency. This subjectivity and lack of transparency has implications for the way that rationality is applied to local government affairs and for how the public participates in government decision-making.

A prototypical rational planning process begins with the collection of data, proceeds to a statement of conditions, involves a public process to identify goals and objectives, and then generates a series of activities to achieve those goals and objectives. Trenton's land-use planning under the leadership of Allan Mallach closely resembled that process, but its execution was weak. In the other cities studied, HI-TOADS were prioritized with little thought to a rational planning process. The result was that HI-TOADS were targeted by cities with little consideration of wider, public goals and objectives.

I recommend a needs-based analysis be conducted by cities to help develop more-objective bases for prioritization of HI-TOADS. Despite its problems, the EPA's Hazard Ranking System (HRS) is a conceptual model for how to prioritize contaminated sites for remediation funding. The HRS compiles extensive data on air, water, and soil pollution at a site, considers the number of residents living in close proximity, and their risk of exposure to contaminants. Then, the HRS generates a score for each site and prioritizes the list of sites for funding and EPA attention in the National Priority List.

A National Research Council (1994) study pointed out several deficiencies in the HRS. The first version of the HRS did a poor job at considering relative risk to human health and the environment. Improvements were made in the 1990 revised HRS. But with those revisions, the HRS became more convoluted and harder to comprehend (National Research Council 1994). Also, the HRS did not weigh the costs and timing of remediation, issues that play critical roles in actual clean-ups.

Despite these weaknesses, the HRS model offers a solid conceptual basis for a similar ranking tool for HI-TOADS. Such a tool could measure actual neighborhoodwide impacts, assess reuse potential, and calculate environmental risk. The sites in each city could then be ranked and prioritized for funding and attention by city, state, and federal agencies.

Political Realities of Policy Recommendations

In the typical land-use dispute, various sides are pitted against each other. Whether it is an issue of siting a new waste-transfer station, converting a single-family house to a group home for the mentally ill, or damming a river, the drawing of battle lines is a perennial problem. In the new-construction arena it is pro-growth versus no-growth groups, in the river-damming arena it is pro-environment versus pro-energy groups. Unique to the HI-TOADS phenomenon is the lack of such clear battle lines. There are no political interest groups that stand to benefit from keeping a derelict site derelict. Everyone would like something to happen, whether it is a priority or not does vary. The conflicts arise on the question of what should be done. Such was the case in deliberations over reuse plans for the Hazel-wood LTV site in Pittsburgh. Smaller, neighborhood-based associations are fearful that powerful philanthropic organizations in the city are not going to consider their concerns in the reuse.

Conflicts also arise over what to do with buildings on a property. Activist historic preservation groups in New Bedford derailed redevelopment efforts at Fairhaven Mills over disputes about how historically significant structures would be preserved.

But when it comes to the question of whether city resources (time, energy, or money) should be spent at a given HI-TOAD site, little formal resistance occurred in cast-study sites. The only resistance comes from within the city as political leaders prioritize other projects over HI-TOADS.[4] With the city government as the only impediment to HI-TOADS reuse, the question becomes: How politically realistic are the aforementioned policy recommendations to better open up public participation, to put more decision-making power in the hands of local residents or NGOs, and invest the city more directly in managing orphaned sites?

To the extent that HI-TOADS are viewed by the electorate as a serious problem not being addressed sufficiently by city government, there will be recognition that making public redevelopment more transparent will be beneficial. Once recognized, these recommendations for greater openness will be politically viable

Just as greater transparency may be rejected by city leaders in the absence of resident demands, officials may feel threatened by greater empowerment of NGOs. However, the Trenton story of Magic Marker illustrates that in the reuse of HI-TOADS there is room to spread credit for good work wide and far. By placing a high level of responsibility into the hands of residents and community activists, the city took a gamble with community empowerment. But the city has received great acclaim for their efforts from the media, the state, the federal government, and independent organizations.

This experience suggests that enhancement of community empowerment funding and authority could be welcome by those familiar with success stories like Magic Marker.

At each stage in the research, I encountered limitations with data. During the Stage 1 analysis, I discovered that the vacant land data from the Bowman and Pagano (2004) survey were unreliable. This was unfortunate, but no other land-use data is available for vacant and abandoned property that is superior to the Pagano data.

During the second stage of the study (telephone interviews with local officials), I encountered further data limitations. While my intention has been to study local officials' activities broadly, I also have been particularly interested in the work of planners. However, in many cities, planners have little or no knowledge of their government's efforts with respect to HI-TOADS. The method that I employed, while it yielded valuable data, involved interviews with some local officials either peripherally or completely uninvolved in HI-TOADS efforts. Nevertheless, it kept the focus on planners and the practice of planning.

A second limitation of the second stage was the absence of many very large cities in the group of 21 cities for which I conducted interviews. From the 21 cities, Baltimore was the largest, with a 2000 population of 651,154. Officials in other large cities, like Detroit (2000 population, 951,270) and Philadelphia (2000 population, 1,517,550), declined participation. This limits the generalizability of the study to medium and large cities.

For the third stage of the research, the case studies, I analyzed property value and socio-economic data for each city. While I was planning to study neighborhood change at the block group level, the computer package I used, Geolytics, did not have 1980 or 1990 census data at the block group level for Census Long Form variables, like housing values or gross rent. Instead, I used census tract data to approximate neighborhoods in each city. Because I was interested in measuring success by the reduction of externalities of abandoned property, the combination of qualitative evidence from interviews and direct observation and the coarse census data was sufficient.

The findings presented in the previous section are important, but need to be considered in light of the limitations of the research. The initial statistical analysis stage used national data and multiple variables to arrive at a list of U.S. cities relatively likely to have neighborhoods with HI-TOADS. The data and variables I used were based largely on dated information from the 2000 U.S. Census. Future research, using data from either the American Housing Survey (for limited metropolitan areas) or the next decennial census could improve on this effort.

The telephone interview stage included 38 interviews with officials in 21 cities. The results proved to be interesting, but future research should attempt to increase the sample size. A possible alternative to the telephone interviews is the distribution of a mail or Internet survey to local officials.

While the sampling frame was limited to cities with populations greater than 100,000 in 1970, future research may seek to interview officials in smaller cities. Smaller cities may address HI-TOADS differently than medium and large cities. For that matter, a narrow focus just on very large cities (of which few were part of this study) could be beneficial.

With respect to the case studies, this research is limited by the sheer number of case studies conducted. Future research could build on this research by extending the number of case studies. The limitation of only studying cities in the Northeast, South, and Midwest could be overcome with the addition of case-study cities in the West.

The case studies were also limited by the narrow time horizon examined, 1995 to 2005. Future research could take more of a historical perspective and study the efforts of local governments to address HI-TOADS over the last twenty-five or fifty years. A similarity emerged in the case studies between city activities relative to HI-TOADS in the last ten years and the urban-renewal efforts of the 1950s and 1960s. These parallels could be studied further through historical research of neighborhoods and their HI-TOADS.

The case-studies analysis was also weakened by my reliance on housing values and rent data from the census in order to study impacts of HI-TOADS reuse. Local assessors in many large cities have computerized appraisal data for their cities at much finer geographies and more regular time intervals than the census. While obtaining such data was cost prohibitive, future research should consider including assessor data in examining the effects of HI-TOADS reuse on neighborhood housing values.

While the study yielded some new information about the role of the private, for-profit sector, those actors were not a focus of the study. Future research could examine more closely the role of the real estate development community in addressing HI-TOADS. Missing from this study was a solid understanding of why and how profit-driven investors can benefit from HI-TOADS projects. Future research could tackle these questions and attempt to determine which developers might be best suited for redeveloping HI-TOADS.

The policy recommendations above are meant to generate discussion in communities with HI-TOADS. For in all cities, in all neighborhoods, there is no one-size-fits-all solution to HI-TOADS. But the tools and techniques described in this study can be useful to local officials and adapted to their needs.

For many of the cities I studied, the past was alive and not well. Officials did not celebrate the past as one might celebrate it in a museum. Except for those in Richmond, many participants saw their cities' once-mighty history as a hindrance. The great pottery mills of Trenton sit empty. The vast acreage of Pittsburgh's steel industry lays fallow. The cotton mills of New Bedford are shuttered. How do cities manage an obsolete landscape of HI-TOADS? If they are smart, they envision a new landscape that fits into the old. If they are smart, they employ the planning tools, techniques, and expertise to create that vision. Rebuilding cities is expensive and difficult. The reuse of HI-TOADS, if done well, can be an important anchor in this new hybrid landscape of old and new.

Epilogue

When I began this research, many officials I spoke to had never heard of HI-TOADS. By the end of most of the interviews, we achieved a sense of shared understanding. They got me. They understood why I was so passionate about that old mill behind the train tracks or that factory that almost burned down last month. My probing often led to more probing. Why are you working on Ohio Works and not the Aeroquip site? How can you preserve the historic resources at Elco Dress Factory while satisfying nearby neighbors who think it's a fire hazard? With all of Mayor Palmer's talk about sustainable development, why are five of the city's eight HI-TOADS still vacant?

As we explored together the answers to these and other questions, a larger story unfolded. The people I met were as passionate as I was, if not more so. They loved their cities. Every day, they worked tirelessly to make them better. Even in small-government Richmond, city officials sat at the edge of their seats, with wide eyes, telling me about the next great thing happening in Shockoe Bottom or Manchester.

As America returns to its cities, urban repopulation is creating a new energy in many neighborhoods to address HI-TOADS. While not as evident in Youngstown and New Bedford, gentrification is occurring throughout Trenton, Pittsburgh, Richmond, and scores of other cities. New residents bring wealth and power to urban neighborhoods and are increasingly demanding that city governments take an active role in redeveloping HI-TOADS.

Where population growth is elusive, a new model appears fitting for places like Youngstown and others: smart decline. Youngstown's 2010 Plan and its recent implementation has garnered national and international attention as part of a wider shrinking-cities movement. Rather than work to reverse population decline, here city officials work to make their city better no matter how it changes. This new model requires that we move beyond the growth/decline dichotomy. Rather, we should focus on managing change in communities and pay less attention to controlling that change. With this new outlook, Youngstown's future does indeed look bright. As the city takes charge of what it can control (such as demolishing abandoned buildings, installing landscaping, and enforcing building codes), it may continue to change demographically, but not in character and spirit. Smart decline provides a new lens by which a city can view HI-TOADS or any other redevelopment challenge. The derelict old factory along Crab Creek is no longer simply a commodity to be traded, but part of a dynamic landscape whose next reuse is unbounded. Within the context of death and life, the smart decline model lets us see neighborhoods as neither alive nor dead but as always in flux—with the job of public policy and planning to try to keep up.

For those cities that are growing, we can tell a more familiar tale of revival and rebirth. As empty-nesters look to retire, single workers look for the thrill and excitement of city living, and families look for affordable communities with a plethora of amenities, these depressed yet growing cities offer potential. Unlike Chicago, New York, Los Angeles, and other global cities, second- and third-tier cities offer low real estate values and potentially high quality of life. The stories in this book give insights into the workings of both governmental and nongovernmental involvement in urban redevelopment in these beginning years of the twenty-first century. The stories are largely positive and give rise to optimism. Cities are mostly engaged in addressing their most-contaminated abandoned sites and that is a good thing. While they are not always successful, there are reasons for failure. With each mistake or obstacle, activists and officials take a turn and reorient themselves. They continue to push on and are tenacious in this important fight.

Appendixes

Detailed Methodology for Identifying Cities for Inclusion in the Study

Abandoned Buildings Dataset, Methods, and Results

In 1997 and 1998, Bowman and Pagano (2004) mailed a self-administered survey to planning officials in all U.S. cities with populations greater than 25,000 in 1995. Their survey sought information about amount of vacant land and number of abandoned buildings in each city. This was the first national survey of abandonment and vacant land since the 1960s. Unfortunately, the survey had quite a low overall response rate, 35 percent. But the response rate for cities with populations greater than 100,000 was a more respectable 50.25 percent. As a first step in trying to answer research question number 1, I turned to these survey results because the presence of abandoned housing and vacant land has been shown to be correlated highly with the presence of HI-TOADS (Greenberg et al. 2000). Many of the same market, environmental, or socio-economic forces that are known to generate HI-TOADS are also known to generate building abandonment and vacant land (Bowman and Pagano 2004).

While initially I was hopeful that the vacant-land data would be of use in this analysis, the definition that Pagano and Bowman provided in their instructions was so broad that the survey results for vacant land are not useful for my purposes. Included in Pagano and Bowman's definition was raw, undeveloped land in a city's outskirts, hardly a predictor of HI-TOADS. Therefore, I only examined the abandoned-buildings data. Eighty-nine cities with populations greater than 100,000 in 1995 responded to the survey and 66 answered questions in the survey concerning abandoned buildings. The sample is representative of the total population of U.S. cities with populations greater than 100,000. The sample includes 33 percent southern cities and 35 percent western cities, where the 2000 census reports that southern and western cities account for 35 percent and 37 percent of all cities with populations greater than 100,000, respectively. The mean number of abandoned buildings in the cities is 1,857. The median is only 162 abandoned buildings, indicating a very strong negative skew. Even by dividing the number of abandoned buildings by the population in each city, the results are still skewed: 3.01 abandoned buildings per 1,000 inhabitants with a standard deviation of 5.86. There is a large range in abandonment, from zero reported abandoned buildings in Irvine, Salinas, and San Jose, California, and Alexandria, Virginia, to 54,000 reported abandoned buildings in Philadelphia, Pennsylvania.

Despite the lack of normality, the prima facie results appear to be fairly consistent with prior research on abandonment. According to Cohen (2001), the City of Baltimore's own survey found 12,700 abandoned housing units, where city officials responded to Pagano and Bowman that there were 15,000 abandoned structures in the city. Likewise, Keating and Sjoquist (2001) cite city officials in 2001 as enumerating 10,000 abandoned buildings in Detroit, where officials there reported 15,000 to Pagano and Bowman.

The ranking of the 66 cities in terms of per capita abandonment is presented in Appendix B. Even though 45 of the original 66 cities (68 percent) are in the South and West, only 55 percent of the top twenty ranked in terms of greatest per capita abandonment are in the South and West. With the exception of Richmond, Virginia, the top-ranked cities, Philadelphia, Baltimore, and Detroit, are all in the traditional manufacturing belt.

U.S. Census Dataset and Methods

Where the first dataset is only a small sample of all large U.S. cities, the U.S. Census Bureau collects relevant socio-economic and housing data for all cities on a decennial basis. I selected U.S. cities with populations of 100,000 or more in 1970, conducted data reduction and multivariate analysis on key variables, ran the analysis again with a larger set of variables, then ended the analysis with a list of the cities in order of their relative likelihood of having neighborhoods with HI-TOADS.

As described earlier, the city size limitation means that my study will not include cities that have only recently been developed heavily, like Irvine, California or Pembroke Pines, Florida (both with greater than 100,000 populations in 1995 and included in the Pagano and Bowman [2004] study). Any city with a population of less than 100,000 in 1970 is unlikely to have neighborhoods with many HI-TOADS. I chose 1970 as the key year because of the deindustrialization that then began to occur in U.S. cities (see chapter 2). However, cities with populations greater than 100,000 in

1970 are today likely to have a number of different neighborhoods, some perhaps having HI-TOADS, others not. The hope is that the 100,000 population limitation on the sample size will improve the likelihood that I identify large cities with many neighborhoods and hence more than one HI-TOAD site.

A total of 153 U.S. cities had populations greater than 100,000 in 1970. These cities are spread fairly evenly throughout the United States, with 45 percent in the Northeast and Midwest and 55 percent in the South and West. The mean population of these cities is 405,980, with 59.5 percent white, and 24.3 percent African American. The median household income is $36,766 and the median housing price $123,090 (see table A.1 for a full socio-economic profile).

Based on a review of the literature presented in chapter 2, I identified eight key variables that helped me identify cities relatively likely to have neighborhoods with HI-TOADS: change in manufacturing employment from 1970 to 2000, median housing value in 2000, median household income in 2000, population change from 1970 to 2000, percentage of vacant housing units, percentage of households receiving public assistance, percentage of housing units without adequate plumbing, and unemployment rate. I described some of these variables in chapter 2. I present further background information on some of the variables in the following section.

Declining manufacturing employment can be viewed as a surrogate for abandoned factories or other obsolete technology facilities. In general, low

Table A.1

Socio-Economic Profile of 153 Cities with Populations Greater Than 100,000 in 1970 and of the Entire United States.

Variable	Mean	25% quartile	75% quartile	Entire United States
Percent 65+	6.1	5.1	6.8	12.4
Percent high school graduate	77.5	73.4	83.1	80.4
Percent receiving public assistance	5.4	3.1	6.7	3.4
Percent without adequate plumbing	1.0	0.6	1.2	1.2
Percent in poverty	17.5	13.7	21.5	12.4
Percent single-mother households	16.1	12.0	19.4	12.2
Percent unemployment	4.7	3.8	5.5	3.7
Percent other vacant housing units	1.6	0.6	2.1	1.5

SOURCE: U.S. Census 2000.

housing values are an indication of the overall strength of the housing market. While HI-TOADS are rarely residential properties, a city with a low mean housing value is likely to have at least some neighborhoods where commercial and industrial demand is also low. Likewise, low median household income indicates that a city is *not* likely to be economically strong but rather a place with poor and distressed neighborhoods (places where HI-TOADS are likely to exist). Cities with high median household income are more likely to have many desirable neighborhoods and less likely to have neighborhoods with HI-TOADS.

A city often loses population after losing a large percentage of its jobs (Bluestone and Harrison 1982). Heavy population decline then indicates that a place is likely to have many former facilities for employment that today may stand vacant. The final variable is the percentage of housing units in a city that were classified as "other" vacant. This "other" category represents unoccupied housing units with the exception of properties for rent or sale or for seasonal, recreational, or occasional use. In a follow-up to the 2000 Census, a study was conducted by the U.S. Department of Housing and Urban Development (2004) examining why housing units were classified as "other" vacant. The study concluded that a majority of units were so classified because they had "fallen into disuse . . . or [were] in transition" (10). This "other" vacant variable will assist in identifying cities with widespread housing abandonment. Widespread abandonment has been shown to be correlated closely with the appearance of HI-TOADS (Greenberg et al. 2000).

In chapter 2, I refer to an extensive body of research that leads me to think that poverty is related to HI-TOADS. I added three poverty-related census variables to the analysis to compare the results with what I found using the land-use and economic variables described above. Those additional variables are percentage of households receiving public assistance, percentage of housing units without complete plumbing facilities, and unemployment rate. All of those variables were collected in the 2000 U.S. Census.

Descriptive statistics for all eight variables are presented in table A.2. In examining skewness, kurtosis, and histograms of each variable, I saw the need to make some transformations to improve the variables' normality. These transformations are important because factor and cluster analysis does not function well with variables with non-normal distributions.

Table A.2

Descriptive Statistics of Variables in Analysis

# Variable	Mean	Median	Standard Deviation	Skewness
1 Change in percentage manufacturing employment (1970–2000)	−43.5	−47.7	18.2	1.39
2 Housing value (2000)	$123,089	$96,300	$78,821	2.10
3 Household income (2000)	$36,766	$35,928	$ 8,644	1.63
4 Population change (1970–2000)	19.3	5.0	50.5	1.88
5 Percentage of "other": vacant housing units (2000)	1.6	1.2	1.5	3.11
6 Percentage receiving public assistance (2000)	5.4	4.6	2.9	1.25
7 Percentage without adequate plumbing	1.0	0.8	0.8	5.44
8 Unemployment rate	4.7	4.5	1.4	0.84
N =	153			

SOURCE: All variables are from 2000 U.S. Census.

While I included all eight variables in the analysis, I determined the first five to be the most theoretically relevant and will focus on them. All five of these key variables are left-skewed, with skewness ranging from 0.84 to 5.44. The manufacturing-change and population-change variables are already derivative of other variables and their non-normality was not severe, so I did not transform them. But I did correct the housing value, income, vacant housing units, and percentage without adequate plumbing variables by converting them by log base ten.

U.S. Census Results

Appendix C shows the correlation matrix for the five key variables. Population- and manufacturing-change variables are highly correlated with each other and with housing vacancy, as would be expected. However, they are not highly correlated with income and housing values. Interestingly, income and housing value variables are highly correlated with one another as well as with housing vacancy, and not correlated with population and manufacturing change. At the same time, income and housing value variables are highly correlated with each other and housing vacancy.

I used factor analysis to sort out some of the more complex interrelationships among the variables. In this analysis, I used a variety of options to ensure that the results did not depend on the factor analysis model chosen.

Table A.3
Table of Runs

Run #	1	2	3	4	5
Method	PCA	PAF	Varimax rotated	Oblimin rotated	PAF w/8 variables
Percent cumulative variance explained	85.2	74.9	85.2	85.2	77.3
Number of factors	2	2	2	2	3

First, I ran principal components analysis on the dataset of 153 cities with all eight variables (see table A.3). The result was high communalities for all variables (>.618) and the production of two factors accounting for 85.2 percent of the variance in the data (see table A.4). Next, I ran principal axis factoring, Varimax rotated, and Oblimin rotated. Finally, in order to see if the addition of the three extra variables (public assistance levels, adequate plumbing, and unemployment) would affect my results, I did another principal axis factoring run. The results for each are displayed in table A.3. In the interest of brevity, I will not describe the detailed results of Runs #1, #3, #4, and #5 and instead just present the results of Run #2, which will be used as the basis for developing the 30-city sample.

Run #2: Principal Axis Factoring (PAF)

Unlike Principle Components Analysis, Principal Axis Factoring does not assume total covariance; nevertheless, only one strong factor emerged. I found high communalities when I ran the data using PAF. Extraction levels for all variables were greater than 0.687. Two factors were generated, which explained nearly 75 percent of variance. The first factor, "HI-TOADS Hosts," had the equivalent explanatory power of almost three variables. The variable factor loadings for both the first and second factor are presented in table A.4. All variables had high factor loadings. The change in manufacturing variable had the lowest loading at 0.534.

The directions of the factor loadings were consistent with what I expected. Reading down the HI-TOADS Hosts column in table A.4, the high positive loadings for the first four variables (0.534, 0.743, 0.841, and 0.696) can be read either positively or negatively. Read negatively, the factor emerges as cities with declines in manufacturing employment, low housing values, low household incomes, and declining populations all align a single factor with the fifth variable: housing vacancy. Where low housing values help us to identify cities likely to host HI-TOADS, low housing vacancy levels do not. Therefore, that variable must be read inversely to the other four to see that it, too, aligns along the same factor.

The second factor had moderate to high factor loadings on only three of the variables and accounted for a mere 18 percent of the variance in the data. I refer to this as the "Built-out Employment Centers" factor, and interpret it in the inverse. Cities with high housing values, stagnant population growth, and growth in manufacturing employment over the last three decades have low factor scores (including Stamford, Connecticut, Cambridge, Massachusetts, and Honolulu, Hawaii). Because these cities are fully built-out, they are not experiencing the kind of explosive population growth that other cities with comparable economic expansion are. These Built-out Employment Centers are not likely to have neighborhoods with HI-TOADS. On the other hand, the first factor, HI-TOADS Hosts, list Camden, Buffalo, and Gary with the highest factor scores, cities well known to have severe abandoned property problems.

I changed the Eigenvalue from its default value of 1.0 to 1.25 to see the effect it would have on the results of the PAF run. As expected, it reduced the number of factors to just one. That one factor had slightly less explanatory power (2.667 sum of squared loadings) as compared to what I found with the Eigenvalue set at 1.0 (2.807 sum of squared loadings). Also, the factor scores for that single factor decreased slightly for three of the four variables; the greatest drop

Table A.4

Factor Loadings for Run #2

	Factor	
		Built-out
Variable	HI-TOADS Hosts	Employment Centers
Variable	1	2
Change in percentage		
manufacturing employment (1970–2000)	0.534	0.641
Housing value	0.743	-0.495
Household income	0.841	-0.279
Population change (1970–2000)	0.696	0.45
Percentage of "other" vacant housing units	-0.882	0.06

SOURCE: All variables are from 2000 U.S. Census except where noted.

was in the manufacturing-change variable, which fell from a factor loading of 0.534 to 0.449. These small differences show that the data is not very sensitive to minor changes in the Eigenvalue, indicating robustness of the results.

Results of Cluster Analysis

In order to check the results of my factor analysis, I ran cluster analyses on the factor scores generated by Run #2. Using the factor scores, I ran the data by clustering observations (cities) to test whether the factor scores generated were robust. As is common practice in cluster analysis, I set the cluster membership range from five to nine clusters.

In each of the five to nine clusters generated, a consistent pattern emerged that put the top sixty cities (identified in the factor analysis) in a single cluster (see Appendix D), with the one exception that Camden was put into its own cluster when cluster memberships were six, seven, eight, and nine. With a cluster membership of five, Camden was part of this single cluster of sixty cities.

Clusters three and four each included at least a dozen cities. Cluster three included several large cities with little in the way of defunct industrial land, including Washington, D.C., Boston, Denver, and Oakland, California. Cluster four included smaller southern and western cities with high to moderate growth and little in the way of HI-TOADS problems. Cluster four included Montgomery, Alabama, Oklahoma City, Oklahoma, Corpus Christi, Texas, and Columbus, Georgia.

The List of Cities Based on Factor Scores for Run #2

Between the five factor analysis runs and the cluster analysis, I am now comfortable moving forward with a close examination of the factor scores to aid in ranking the cities in terms of their relative likelihood to have neighborhoods with HI-TOADS. I selected the results of Run #2 to generate the ranking due to the superiority of principal axis factoring for this type of analysis and the robustness of that run. Appendix B features a list of the 153 cities with their census ranking indicated in the third column.

In the research for this book, I followed this analysis with a series of interviews with local officials. Therefore, the goal of this analysis was to generate a sample of 30 cities that had the highest relative likelihood of having

Table A.5

Geographic Distribution of 153 Cities in Population and 30 Cities in Sample

Region	Population of cities		Sample of cities	
	Frequency	*Percent (%)*	*Frequency*	*Percent (%)*
Northeast	28	18.3	13	43.3
Midwest	41	26.8	9	30.0
South	52	34.0	8	26.7
West	32	20.9	0	0.0
N =	153	100	30	100

neighborhoods with HI-TOADS. Table A.5 shows the geographic distribution of the thirty top-ranked cities in comparison to the distribution of all 153 cities. The geographic distribution of the thirty cities and the 153 cities is presented in Appendix E. While the study sample has a strong bias toward the Northeast and Midwest (accounting for 73.3 percent of the sample), the South is decently represented (26.7 percent of the sample). Noticeably absent from the sample are any western cities.

For each of the socio-economic and housing variables examined, the sample is statistically significantly different from the population at a minimum of 95 percent confidence level (the sample is statistically significantly different at the 99.99 percent confidence level for all but two variables, percent older than 65 and population; see table A.6). Because of the way that I created the sample of cities, I would expect this of a group of cities with lower housing values, lower income, declining population, and declining manufacturing employment. What is striking about the sample group is the extent to which they are so much smaller (30 percent lower population), ensconced in poverty (37 percent higher poverty rate), and populated by African-American residents (45.2 percent for the sample versus 24.3 percent for the population) than the 153 cities.

Aggregating the Lists

Both the census and abandoned housing datasets offer valuable insights into which U.S. cities are relatively likely to have neighborhoods with HI-TOADS. In Appendix B, the fourth column includes the rank order of the cities included in the abandoned housing dataset and also included in the census analysis. The 35 overlapping cities are presented in Appendix B

Table A.6

Comparison between 153 Cities in Population and 30 Cities in Sample

Variable (from 2000 Census)	Population of Cities (Mean)	Sample of Cities (Mean)
Population	405,980	281,299
Median housing value	$ 123,090	$ 71,587
Median household income	$36,766	$27,962
Percent 65+	6.1	6.3
Percent African American	24.3	45.2
Percent white	59.6	45.6
Percent high school graduate	77.5	72.5
Percent receiving public assistance	5.4	8.2
Percent without adequate plumbing	1.0	1.7
Percent in poverty	17.5	24.0
Percent single-mother households	16.1	23.5
Percent unemployment	4.7	6.3
Percent other vacant housing units	1.6	3.7
N=	153	30

All sample variables are statistically significantly different from population at the 99.99 percent confidence level, except Population and Older People (95 and 99 percent confidence levels).

alongside their census ranking. At first glance, there appears to be a relationship between cities that ranked high on abandonment (from the Bowman and Pagano survey results) and those that ranked high on the census ranking. An ordinal rank correlation reveals a Spearman's rho correlation of 0.621 (significant at the 0.01 level) between the two rankings. This moderate correlation suggests the robustness of the census ranking and is sufficient for me to move ahead with Appendix B as the final list of cities.

Protocol Employed to Secure Interviews:

1. I placed phone calls to the local official identified as the point of contact and left at least two messages;
2. if the local official returned a call and left me a message, I made more calls and left at least two more messages;
3. if the official did not respond to the calls, I sent an email message (in cases where email addresses were available);

4. after conducting the interview with the first local official, I sought a referral to another official;
5. I placed phone calls to the second local official and left at least three messages;
6. if the second official returned a call, I made more calls and left at least two more messages;
7. if the official did not respond to the calls, I sent an email message (in cases where email addresses were available).

Ranking of U.S. Cities on
Likelihood of Hosting HI-TOADS Using Five Key Variables

City	State	Census	Abandoned	City	State	Census	Abandoned
Camden	NJ	1	N/A	Allentown	PA	37	N/A
Buffalo	NY	2	N/A	Springfield	MA	38	N/A
Gary	IN	3	N/A	Jackson	MS	39	N/A
St. Louis	MO	4	N/A	Toledo	OH	40	N/A
Pittsburgh	PA	5	N/A	Milwaukee	WI	41	N/A
Youngstown	OH	6	N/A	Waterbury	CT	42	N/A
Baltimore	MD	7	2	Shreveport	LA	43	N/A
Philadelphia	PA	8	1	Portsmouth	VA	44	N/A
Trenton	NJ	9	N/A	Hammond	IN	45	N/A
Flint	MI	10	N/A	Mobile	AL	46	6
Dayton	OH	11	N/A	Akron	OH	47	29
Cleveland	OH	12	N/A	Memphis	TN	48	N/A
Birmingham	AL	13	N/A	Bridgeport	CT	49	32
Rochester	NY	14	N/A	Tampa	FL	50	N/A
Syracuse	NY	15	17	Kansas City	MO	51	N/A
New Orleans	LA	16	N/A	South Bend	IN	52	12
Detroit	MI	17	4	Chattanooga	TN	53	N/A
Macon	GA	18	N/A	Duluth	MN	54	N/A
Scranton	PA	19	N/A	Topeka	KS	55	N/A
Hartford	CT	20	N/A	Lansing	MI	56	N/A
Louisville	KY	21	9	Baton Rouge	LA	57	N/A
Albany	NY	22	N/A	Columbia	SC	58	34
Newark	NJ	23	N/A	Norfolk	VA	59	N/A
Savannah	GA	24	N/A	Atlanta	GA	60	N/A
Erie	PA	25	23	Des Moines	IA	61	N/A
Cincinnati	OH	26	19	Washington	DC	62	8
Richmond	VA	27	3	Peoria	IL	63	N/A
Kansas City	KS	28	5	Springfield	MO	64	7
New Haven	CT	29	14	Chicago	IL	65	28
Knoxville	TN	30	39	Independence	MO	66	N/A
New Bedford	MA	31	N/A	Fort Wayne	IN	67	N/A
Canton	OH	32	N/A	Lubbock	TX	68	N/A
Miami	FL	33	N/A	Indianapolis	IN	69	N/A
Evansville	IN	34	N/A	Little Rock	AR	70	16
Beaumont	TX	35	13	Jersey City	NJ	71	N/A
Providence	RI	36	11	Tulsa	OK	72	N/A

City	State	Census	Abandoned	City	State	Census	Abandoned
Montgomery	AL	73	N/A	Warren	MI	114	N/A
Rockford	IL	74	N/A	Tucson	AZ	115	N/A
Elizabeth	NJ	75	N/A	Los Angeles	CA	116	N/A
Worcester	MA	76	41	Yonkers	NY	117	N/A
Oklahoma City	OK	77	N/A	Sacramento	CA	118	N/A
Corpus Christi	TX	78	N/A	Fresno	CA	119	N/A
Columbus	GA	79	N/A	Honolulu	HI	120	N/A
Columbus	OH	80	26	Berkeley	CA	121	N/A
Fort Worth	TX	81	N/A	Long Beach	CA	122	N/A
Jacksonville	FL	82	15	Hialeah	FL	123	N/A
St. Petersburgh	FL	83	N/A	Seattle	WA	124	N/A
Paterson	NJ	84	N/A	Portland	OR	125	N/A
Houston	TX	85	N/A	Lexington	KY	126	N/A
San Antonio	TX	86	20	Pasadena	CA	127	N/A
Grand Rapids	MI	87	N/A	Phoenix	AZ	128	N/A
Newport News	VA	88	33	Lincoln	NE	129	N/A
Minneapolis	MN	89	N/A	Cambridge	MA	130	N/A
Boston	MA	90	N/A	Stockton	CA	131	37
Dallas	TX	91	N/A	Madison	WI	132	N/A
Winston-Salem	NC	92	N/A	Albuquerque	NM	133	N/A
New York	NY	93	N/A	Riverside	CA	134	N/A
Nashville-				Charlotte	NC	135	22
Davidson	TN	94	N/A	San Francisco	CA	136	N/A
Spokane	WA	95	N/A	Glendale	CA	137	36
Omaha	NE	96	N/A	Stamford	CT	138	N/A
Cedar Rapids	IA	97	N/A	Alexandria	VA	139	N/A
El Paso	TX	98	N/A	San Diego	CA	140	45
Hampton	VA	99	N/A	Garden Grove	CA	141	N/A
Hollywood	FL	100	N/A	Torrance	CA	142	N/A
Amarillo	TX	101	30	Raleigh	NC	143	N/A
Fort Lauderdale	FL	102	N/A	Santa Ana	CA	144	N/A
Tacoma	WA	103	N/A	Livonia	MI	145	N/A
Hunstville	AL	104	N/A	Virginia Beach	VA	146	25
St. Paul	MN	105	31	Anaheim	CA	147	N/A
San Bernardino	CA	106	N/A	Colorado Springs	CO	148	N/A
Greensboro	NC	107	N/A	Austin	TX	149	N/A
Wichita	KS	108	N/A	Las Vegas	NV	150	N/A
Denver	CO	109	N/A	Huntington			
Parma	OH	110	N/A	Beach	CA	151	N/A
Oakland	CA	111	N/A	San Jose	CA	152	N/A
Dearborn	MI	112	N/A	Fremont	CA	153	N/A
Salt Lake City	UT	113	18				

Correlation Matrix for Five Key Variables

	Manufacturing change	House value	Income	Population change	Vacancy
Correlation					
Change in percentage manu- facturing employment (1970–2000)	1.000	0.071	0.288	0.661	-0.423
Housing value	0.071	1.000	0.765	0.307	-0.678
Household income	0.288	0.765	1.000	0.438	-0.764
Population change (1970–2000)	0.661	0.307	0.438	1.000	-0.595
Percentage of "other" vacant housing units	-0.423	-0.678	-0.764	-0.595	1.000
Sig. (1-tailed)					
Change in percentage manu- facturing employment (1970–2000)		0.192	0.000	0.000	0.000
Housing value	0.192		0.000	0.000	0.000
Household income	0.000	0.000		0.000	0.000
Population change (1970–2000)	0.000	0.000	0.000		0.000
Percentage of "other" vacant housing units	0.000	0.000	0.000	0.000	

Results of Cluster Analysis Using a Setting of Five Clusters

Cluster 1

Camden	Detroit	Miami	Bridgeport
Buffalo	Macon	Evansville	Tampa
Gary	Scranton	Beaumont	Kansas City
St. Louis	Hartford	Providence	South Bend
Pittsburgh	Louisville	Allentown	Chattanooga
Youngstown	Albany	Springfield	Duluth
Baltimore	Newark	Jackson	Topeka
Philadelphia	Savannah	Toledo	Lansing
Trenton	Erie	Milwaukee	Baton Rouge
Flint	Cincinnati	Waterbury	Columbia
Dayton	Richmond	Shreveport	Norfolk
Cleveland	Kansas City	Portsmouth	Atlanta
Birmingham	New Haven	Hammond	Des Moines
Rochester	Knoxville	Mobile	
Syracuse	New Bedford	Akron	
New Orleans	Canton	Memphis	

Cluster 2

Washington	Oakland	Long Beach	Stamford
Boston	Dearborn	Seattle	Alexandria
New York	Warren	Cambridge	Torrance
Denver	Honolulu	San Francisco	Huntington Beach
Parma	Berkeley	Glendale	

Cluster 3

Montgomery	Amarillo	Phoenix	Garden Grove
Oklahoma City	San Bernardino	Lincoln	Raleigh
Corpus Christi	Wichita	Stockton	Santa Ana
Columbus	Sacramento	Madison	Virginia Beach
St. Petersburgh	Fresno	Albuquerque	Anaheim
San Antonio	Hialeah	Riverside	Colorado Springs
Spokane	Portland	Charlotte	
El Paso	Lexington	San Diego	

Cluster 4

Austin   Las Vegas

Cluster 5

San Jose   Fremont

U.S. Cities with Populations Greater than 100,000 in 1970

LEGEND

● Cities selected for sample

· Cities included in statistical analysis

200　0　200　400　Miles

Interview Instrument

Edward J. Bloustein School of Planning & Public Policy
Rutgers University
Abandoned Properties Study
September 2005

Questions

1) What is your current position? How long have you been in that position, how long have you worked for the City of X?

2) The focus of this research is on High-Impact Temporarily Obsolete Abandoned Derelict Sites (HI-TOADS):

   • Formerly developed, but presently unutilized, abandoned real property which has a measurable impact on property values more than ¼ mile away.

   Are there any HI-TOADS in City of X? Which neighborhoods are they located in? How many?

3) Do HI-TOADS represent a threat to the stability of any neighborhood in City of X?

4) Beyond property values, do HI-TOADS represent any other threats in your City of X?

5) Do you see a role for the City government in addressing HI-TOADS? Why or why not?

6) Has the City government actively tried to address any HI-TOADS? Why some and not others?

7) What policies has the City adopted to address HI-TOADS?

   • developer incentives to encourage redevelopment?

   • eminent domain authority to assemble large parcels?

   • use of zoning and subdivision control regulations?

8) For the HI-TOADS you are working on, what kinds of new uses are being explored?

9) How successful are the policies you have used?

10) Have there been measurable fiscal, economic, social, quality of life, environmental, or aesthetic impacts due to the City's intervention?

11) Has the city's intervention arrested the spread of blight?

Detailed List of HI-TOADS

For each city, between one and seven HI-TOADS are listed.
If available, the following information is presented: site name, size, location, and current status.

City	1	2	3	4	5	6	7
Baltimore, Maryland	Proctor and Gamble Soap Manufacturing Facility, 30 acres (South Baltimore Peninsula). City helped to finance redevelopment into office space in 2000 that provides jobs for 1,600 people.	Montgomery Ward Catalog Distribution Center, 1.3 million square feet, vacated in mid-1980s (1800 Washington Boulevard). With aid of city financial incentives, the site was redeveloped recently into office space.	Rosemont Industrial Area. Agglomeration of a few dozen largely abandoned industrial operators surrounded by a residential area (Lafayette Street area in West Baltimore).	Armco Steel Plant, 40± acres (Rolling Mill Road, East Baltimore). City will not acquire the site, prepared to offer a wide range of incentives to encourage redevelopment. Currently, reuse is held up in litigation.			

City	1	2	3	4	5	6	7
Buffalo, New York	National Fuel Gas Site, heavily contaminated former gas plant, 7 acres (Downtown: West Genesee and 7th Street). City worked with property owner and worked out a deal to remediate the site and build a 490K-square-foot corporate headquarters on the site.	South Buffalo Industrial Area, former pig-iron and steel plants, 1,200 acres. City was awarded a Brownfields Opportunity Area planning grant from the State of New York.	Hanna Furnace, steel production, 113 acres (South Buffalo).				
Camden, New Jersey	Camden Waterfront, roughly 100 acres (along the river). Mixed use entertainment, tourism, residential, park development	Victor Building, former RCA Plant, 4± acres (Front St / Cooper St). With the aid of a number of public agencies, the plant was redeveloped into housing.	Consolidated Foam Site, aka Borden Chemical, 5± acres (Federal Street). City is working with U.S. EPA to demolish structures.	Galliger Site (Broadway and Atlantic in S. Camden. Cleared site, empty and awaiting developer.	Citka Site (Broadway and Atlantic in S. Camden, just West of Galliger Site). Cleared site, empty and awaiting developer.	Old Industrial Belt, string of abandoned factories and lumber yards, 30± acres (US30 / 17th St.).	Gateway Industrial Area, ring of former industrial sites surrounding Campbell Soup World HQ, 30 ± acres (S.11th / Mt. Vernon Ave.).

Cincinnati, Ohio	Affordable Housing Complex, 500 units, abandoned (Paddock St. / Seymour St). It was torn down and developer is building 150 market-rate homes.	Mill Creek Corridor (along Mill Creek).	Ohio River Corridor	Queen City Barrel.	
Dayton, Ohio	Fridgidare Site (Webster's Station). City has control of the property and is getting it shovel ready.	Two RR freight warehouses (Webster's Station), abandoned for 30–40 years. Converted into a public market.	Techtown (Webster's Station). Acquired property, got it shovel ready and is now being redeveloped.		
Gary, Indiana	Buffington Harbor, 50 acres (along the river). Was redeveloped, no longer a HI-TOAD.	Gary Screw and Bolt, 10–15 acres (Rhode Island Street near I-65). Acquired by now-defunct quasi-public agency. Reuse is mired in litigation surrounding the quasi-public agency and its assets.	Dean Mitchell Powerplant, 10± acres (near Lakefront). City controls property, working on passive/low-intensity reuse plans.	Bear Brand Hosiery Factory, 5–7 acres (2100 block Massachusetts Street).	Closed Gary Sanitary Landfill, 100 acres (19th/Burr Street). Recreational uses are being planned during on-site remediation.

City	1	2	3	4	5	6	7
Hartford, Connecticut	Colt Firearm Plant, partially abandoned complex of 12 buildings on 20± acres (Van Dike St./ Huyshope Ave.). $110 million public-private effort commenced in 2004 to adaptively reuse much of the historic site and establish a national park.	Mortson Street and Putnam Heights, a block of abandoned six-family homes, 4± acres. City sold tax liens on the homes to facilitate site control, then brought in public monies and worked with developer to clean the buildings of asbestos and lead-based paint and rehabbed them.	Cliny Street Manufacturing Plant, structures have been demolished but in need of environmental remediation, 1.5 acres (Cliny Street). City is involved in ongoing environmental investigation/ remediation.	Capewell Manufacturing Site, former horseshoe nail factory, 6 acres (Charter Oak Rd/ Columbus Rd.). City is providing grant funds to support the conversion of the property into condominiums.	Homestead Avenue Corridor, commercial district comprised of old factories, 30± acres (Homestead Ave./ Albany Ave.). City is activity seeking a developer to partner on the remediation and redevelopment of the properties in the corridor. An Urban Renewal Plan is in place for a portion of the corridor.		
Knoxville, Tennessee	Coster Shop Rail-yards.	Marble Works, 50± acres (along Merriville Pike).	Witherspoon Site. Was dumping ground for nuclear waste; future is unclear.	I-275 Business Park. Redeveloped successfully.			

Louisville, Kentucky	Scrap Metal site (in Park Hill Corridor). Mixed-use development happened.	Rodius, 17 acres (11th & Hill) former chemical manufacturer.	Voght Commons (between 9th and 15th Streets on Magnolia). City acquired property, in midst of environmental investi-	Park Hill Corridor (1,300 acres of moribund industrial sites).	30th Street Corridor (industrial properties along rail line).	Waterfront. Waterfront Park was developed.
Macon, Georgia	*No HI-TOADS*					
New Bedford, Massachusetts	Fish Processing Center, 25 acres of unimproved land sat vacant for 75 years (on the Waterfront). Mayor led an effort to acquire the property, obtained state and federal money, and today there are three fish-processing companies in operation.	Aerovox Mill Site, manufactured capacitors and generated vast PCB contamination in the building and in the Harbor, 10.5 acres (740 Belleville Avenue). City, through an LLC, is conducting investigation and remediation.	Morris Cutting Tools Plant, 8± acres (163 Pleasant Street). Awaiting completion of the cleanup in next couple years and then will develop reuse plan in concert with the neighborhood.	The Hicks-Logan-Sawyer District, agglomeration of partially occupied, underutilized mill buildings, 95 acres (bounded by Sawyer St, Worcester St., and Route 18). City is developing a strategic plan for the area and will explore urban renewal action there.		

City	1	2	3	4	5	6	7
New Haven, Connecticut	Winchester Rifle Site, 80 acres (Winchester Avenue). Over the last twenty years, city has been heavily involved in redeveloping it into a technology park.	Bigelow Boiler Factory, 8 acres (198 River Street). Property is in an MDP. City is studying contamination issues.	St. Gobain Plastics Factory, largely abandoned for three years, 7.1 acres (East Street/ Humphrey St). High-profile location, has been privately marketed; city may intervene.	Clock Factory, 1.9 acres, largely abandoned (Hamilton St./ Chapel St.). No action to redevelop.	Vacant Hess Oil Terminal, improved by vacant buildings and empty petroleum tanks, 8.7 acres (River Street). City is in the process of acquiring the property with the intent to demolish the structures, subdivide the property, and market it.	Starter Site, former textile factory, 7± acres (James St / Lombard St). No action to redevelop.	
Newark, New Jersey	Haynes Griffith Building (Downtown), former department store. Redeveloped into 518 residential units, 7,000 square feet of retail, and parking.	1180 Raymond Boulevard (Downtown), former office building. Redeveloped into 317 residential units and retail.	Home Depot, site of several abandoned buildings. Being redeveloped into a Home Depot.	Bayonne Barrel (East End), hazardous waste on site.	Whitehose Chemical (East End), chemicals and dyes.		

Pittsburgh, Pennsylvania	South Side Works, old steel mill site (South Side neighborhood). Redeveloped for mixed use.	Nine Mile Run, former slag heap. Redeveloped for mixed use.	Tech Center (across the river from South Side Works). Redeveloped for offices.	Northern Shore of Allegany River. Redeveloped for mixed use (including sports facilities).	Homewood area. No attention being paid there.	
Richmond, Virginia	Manchester Lofts Building (Hull Street and Manchester). Light industrial uses have reoccupied the structures. Originally a warehouse building and a paper box manufacturer.	Tobacco Row (between Main and Carey Streets). Converted to offices and condos.	Todd's Hams Building. Meat-packing plant (Hermitage Rd and West Lee Street). Redeveloped for residential uses.	Marshall Street Industrial Corridor (between Marshall and Broad, near Virginia Commonwealth University). University has torn down and cleaned industrial sites, and rehabbed buildings for educational uses.	Tredegar Civil War Site, iron maker, shuttered in 1970s. The National Parks Service acquired it and established a tourism promotion office.	Fulton Gas Works (Downtown on the James River). Richmond Cold Storage. Converting into housing, offices.
Rochester, New York	GM Delco Plant (Orchard and Whitney streets).	Abandoned junkyards, 10± acres (Oak Street, between Smith and Lyle). Being redeveloped for a new soccer stadium.				

City	1	2	3	4	5	6	7
Savannah, Georgia	Neal Blun Saw Mill, 2.9 acres (52nd St / Montgomery St). Local college acquired and adaptively reused.	Central Georgia Railroad structures. 10 acres (MLK Blvd/ Harris). Local college acquired and adaptively reused.	Gas Station, 2,000 square feet, sat abandoned for 50 years (Drayton Street and Liberty Street). Local university acquired the property and is in the process of adaptively reusing it.	Saw Mill (President Street/ Old Depot Road). City acquired and demolished the structures. Property now sits vacant, no immediate prospects for redevelopment.	MLK Gas Station (Gwinnet Street / Martin Luther King Jr. Blvd). City has designated the neighborhood as a targeted redevelopment area.		
Scranton, Pennsylvania	Delaware-Lackawanna-Hudson Railyard. New National Park was developed around 1993.	Casey Hotel, ½ acre (Lackawanna Ave. / Adams Ave.). After years of neglect, this abandoned building was torn down and the city assisted in the construction of a parking garage on the site.	Scranton Lace Mill, 10± acres (Green Ridge St./ Providence Road). Local college is planning to acquire and redevelop the property.	Eureka Battery Factory, 3± acres (Sanderson St. / Main St). City's redevelopment authority acquired the property and is developing plans for reuse.			

Syracuse, New York	Carousel Mall Site, 75± acres former junkyard (Lakefront Area Triangle). Redeveloped as a regional mall.	Tank Farm, former industrial area (Lakefront Area Triangle). Fully remediated; city is aiding with redevelopment efforts.	Erie Boulevard/ Rt. 690 Corridor, a number of HI-TOADS in the land between those two roads.			
Trenton, New Jersey	Roebling Complex. Mixed use redevelopment in early 1990s.	Crane Pottery Site, vacant land was cleared for development (North Clinton Avenue). Five light industrial facilities were erected.	Assumpink Greenway, abandoned railyards (near train station). Effort is underway to explore converting the site into a greenway or community center.	Prospect/ Remmington/ Oakland streets, 2 properties.	American Bridge and Steel Site, former metal scrap yard (on Delaware River). County-led effort in 1990s to convert site to three 100,000-square-foot buildings, a stadium, and an entertainment complex.	Magic Marker. Northwest Trenton neighborhood, roughly 5 acres.

City	1	2	3	4	5	6	7
Youngstown, Ohio	Abandoned Steel Mill, 26 acres (East Front St.). Is being redeveloped as Youngstown Convocation Center.	Performance Park, abandoned Republic Steel slag processing yard, 70± acres (Poland Ave.). City helped to redevelop it into a modern industrial park.	Salt Springs Industrial Park, 230± acres (Near Meridian Road). Converted recently from a slag dump into an active industrial park.	Former Republic Hose Site, 12 acres (Albert Street). City acquired site and is the midst of applying for environmental assessment funds.	Building and Materials Supply Center, 12 acres (Logan & Hubbard Avenue). City acquired site and is in the midst of applying for state funding for remediation.		

Case-Study Methodology

For each case-study city, I contacted the two officials ("points of contact") with whom telephone interviews were conducted for the earlier stage of the book. In a letter, I asked these points of contact for their participation in further research on the same topics that were covered in the telephone interviews. I consulted with both points of contact to generate a list of potential interviewees for the case study and to coordinate visits to key HI-TOADS in their city. I used the following protocol for ensuring a wide coverage of interviews across different organization levels and hierarchal levels:

1. city officials beyond the points of contact (current or former);
2. county, state, or federal officials (current or former);
3. employees of nonprofit organizations or local residents involved in HI-TOADS reuse.

I sent hard-copy letters or emails to potential interviewees identified with the points of contact asking them to participate in the research. I made three follow-up phone calls to each potential interviewee to attempt to secure an interview. I conducted at least two interviews in each category. If I could not conduct all the necessary interviews based solely on the list generated with the points of contact, I requested referrals from those who I did interview, using a snowball sampling method. Efforts were made to first conduct all interviews in person. If an in-person interview was not logistically possible, interviews were conducted over the telephone.

I conducted focused interviews, where I utilized a set of questions based loosely around the instrument developed during the telephone interviews (Yin 1994). I also developed a new set of questions aimed at exploring key research questions for each of the case-study cities. I will introduce those questions at the beginning of each case study. In each case-study city, local officials and employees of nonprofit organizations took me on driving tours of each city's HI-TOADS. In addition, I explored several key neighborhoods independently and took extensive photographs and written notes during my visits. During this direct observation of HI-TOADS and their neighborhoods, I paid particular attention to building and site conditions, the extent of security at the site, neighborhood conditions, and indicators of criminal activity, arson, or dumping.

As part of the city visits, I analyzed key documentation. This documentation included planning studies, reports, newspaper clippings, and site maps. In addition, I searched several databases for articles, studies, books, and conference presentations on each case-study city from 1995 to 2006, including Academic Search Premier, Digital Books, World-Wide Political Science Abstracts, the Rutgers University Library, and Avery Architectural Periodicals. Also, I searched the Lexis-Nexis newspaper database for stories from between 1995 and 2006 related to "brownfields" or "redevelopment" for each HI-TOAD site in each of the five case-study cities.

The final component of the case-study methodology was a statistical analysis of property-value data for each city. For each city, I generated GIS maps and created new data layers of all HI-TOADS. The enumeration and location of HI-TOADS was based on the interviews and site visits. Once I plotted the location of each HI-TOAD site, I then determined the site's census block group, tract, and zip code in order to conduct the demographic analysis. Following Greenberg and Hollander (2006), I first looked at decennial census data and did a simple comparison of 1980, 1990, and 2000 data to determine whether housing values and rents changed in each census tract relative to the city as a whole. I chose to use tracts for the time series analysis because they generally are viewed as a good approximation of a neighborhood (Rohe and Stewart 1996). Because of the neighborhood effect of the HI-TOAD site, census tracts provide a way to measure how HI-TOADS neighborhoods may differ. Census blocks are too small and do not contain all the demographic data available at the tract level.

Next, I collected data from the company Trend MLS for two key neighborhoods in one of the case study cities, Trenton. Trend MLS collects and distributes residential real estate sales and transaction data to real estate professionals. They were generous in providing basic information for this project at no cost. They provided information on the number of real estate transactions, the mean sales price, and the number of days of the market for two years: 1998 and 2005 (1998 was the earliest year they had available and 2005 was the most recent). The data was provided at the smallest level of geography they had available, the zip code area.

The combined use of focused interviews, direct observation, document analysis, and statistical analysis of census and real estate sales data aided in triangulating my data to arrive at valid and reliable results. The four different research methods, used in concert, provided an effective way for me to pursue key research questions at each city to arrive at results that could be compared across the cities. While the quantitative data on housing values adds much to each case study, it is only illustrative of the broader trends at work in these communities. I am not attempting to prove a causal link between the existence or redevelopment of HI-TOADS and neighborhood change; to do so would require much more rigorous statistical analysis and more detailed data. The focus of this book is on the qualitative evidence presented in the interviews, derived through analysis of plans and reports, and through direct observation.

As was done for the information presented in chapter 3, research participants in the case-study chapters were asked to sign an Informed Consent Statement and to indicate whether they wished for their anonymity to be protected. While many officials were comfortable with me attributing quotes to them, in the interest of consistency, I do not use any identifiers throughout the case studies.

Notes

1. Introduction (pages 1–12)

1. I am focusing on whether cities are likely to have HI-TOADS and not on any particular number or per capita amount of HI-TOADS.

2. Research, Writing, and Thinking on HI-TOADS (pages 13–29)

1. Examples of relevant federal laws introduced in the twentieth century include the Clean Air Act of 1963, Clean Water Act of 1977, and Comprehensive Environmental Response, Compensation, and Liability Act (CERCLA) of 1980.
2. Accordino and Johnson (2000) used the U.S. Census regional classification system in examining western, midwestern, northeastern, and southeastern cities.
3. These policies also can be used to prevent HI-TOADS. However, in this study, I am primarily interested in how cities address actual HI-TOADS and less on how cities prevent them.
4. In defining a successful reuse effort, this book will concentrate less on resident well-being, and more on the reduction in off-site externalities of abandoned sites (not to the exclusion of consideration of resident well-being).

3. Local Officials and Their Attitudes toward HI-TOADS (pages 30–46)

1. The indicators were: percentage change in manufacturing employment from 1970 to 2000, percentage population change from 1970 to 2000, housing values in 2000, household income in 2000, percentage of "other" vacant housing units in 2000.
2. Efforts were made to conduct interviews with heads of planning or community development departments, as well as with at least one other local official. The second interviewee was an economic development official, brownfields coordinator, or city manager. The selection of the second interviewee was based on a referral from the first point of contact. Who participated in that second interview varied depending on the circumstances of each city, its governmental structure, and the relationships among its agencies.

3. New Orleans was originally one of the 30 cities to be studied, but because of the devastation caused by Hurricane Katrina on August 29, 2005, I removed it from the sample and included New Bedford, Massachusetts (which ranked 31 on the index) instead.

4. I conducted the research at two institutions and began by sending letters on letterhead from Rutgers University and later began teaching at Tufts University and began to use that letterhead.

5. Notes were taken of each interview. I subsequently analyzed the notes thematically, to understand the dimensions of the responses to each question and to understand larger ideas and messages conveyed during the interviews. Because this study involves two discrete units of analysis, cities and officials, I needed to devise a system by which I could calculate and analyze the interview results on both bases. I entered the interview results directly into a spreadsheet, with a row for each observation (each local official). Then, I generated a second spreadsheet where each city was listed as an observation. I based the data for each city upon a comparison among the answers provided by each official at that city. When the answers among the officials differed (for example, one official said that arson was a problem and the other did not), I entered and recorded the response of the more senior (in terms of years in current position) official.

6. Two officials were contacted at the thirty cities that ranked highest on the list of cities most likely to host HI-TOADS. As indicated earlier, I complied with the Institutional Review Board's Human Subjects Review process and received an exemption based upon my research protocol. For each of the interviewees, I requested them to select one of two informed consent agreement forms notifying me either that they would like to have their anonymity protected or that they did not. Of the 38 officials interviewed, 24 completed the informed consent agreement. Of those, six (25 percent) requested that their anonymity be protected. Therefore, in the remainder of the chapter, quotes will be attributed directly to officials who gave permission to do so. Quotes will not be attributed to those 20 officials who either did not complete the informed consent agreement or did so but indicated that they preferred the interview be used anonymously.

7. Interviewees described distribution, warehousing, storage, and junkyard facilities as "low-end" uses.

8. Brownlining is a practice whereby banks restrict their lending exclusively to greenfield properties and avoiding sites with any known or perceived contamination (Yount and Meyer 1994). Brownlining gets its name from the analogous red-lining practice of banks that draw lines around neighborhoods based upon their racial and ethnic composition and restrict lending in those areas.

4. A National Perspective on What Cities Do about HI-TOADS (pages 47–77)

1. I use 20 here because only 20 of the 21 cities reported having current HI-TOADS.

2. Details of some of those redevelopment projects are provided in Appendix G.

3. In 1993, Indiana was one of the first states to adopt a Voluntary Remediation Program, which protects local governments and other nonculpable owners from environmental liability associated with redevelopment projects on contaminated property. The program also protects responsible parties by limiting what otherwise would be unbounded liability for clean-up activities.

4. When state-wide funding requires job creation at a site, then that effectively trumps what local residents might desire for the site, for example open space, recreational facilities, or agriculture.

5. While the planned and envisioned land uses are important to understand, this study has focused primarily on outcomes; what matters most in neighborhoods is what actually gets built.

6. It is important to note that only a single city used zoning and considered itself to be unsuccessful. These results are provocative, yet due to the small number of cases, the validity of these emerging relationships is compromised. These trends will be explored further in the case study chapters.

7. The Spearman's correlation of -0.442 represents a moderate inverse relationship between grant monies and planning implementation.

8. An alternative explanation could be that officials involved in HI-TOADS reuse are well-suited to the task because of their unique understanding of real estate development and finance.

9. See the profile for Oak Street Junkyards in Rochester, New York, earlier in this chapter.

5. Redevelopment Policy in a Municipal Coalition City: Trenton, New Jersey (pages 78–117)

1. The case study research was conducted from January 2006 to December 2006.

2. Of those who reported being of one race (U.S. Census 2000).

3. For owner-occupied housing units. Mean housing value can be calculated by taking the aggregate housing value for an entire geographic area and dividing that amount by the number of housing units.

4. However, this growth in incomes was slower than the growth of inflation from 1980 to 2000, which was 109.0 percent. Failure of incomes to keep pace with inflation can mean a relative decline in consumer purchasing power. Based on a Consumer Price Index of 82.40 in 1980 and 172.20 in 2000, inflation can be calculated using the following formula: $[(172.20-82.40)/(82.40)*100] = 109.0$ percent inflation from 1980 to 2000.

5. This property value appreciation occurred without the introduction of a substantial amount of new housing construction.

6. EPA established a number of placements throughout the country particularly in support of the brownfields program.

7. The Assunpink Greenway Steering Committee has only a single representative from a nongovernmental organization, the Stonybrook Millstone Watershed Association, based in rural Pennington, New Jersey.

6. Slag Heap, Steel Mills, and Sears: Pittsburgh, Pennsylvania (pages 118–48)

1. Butler buildings are constructed of light-weight steel and are designed for temporary uses.
2. The Green Building Alliance recently had hired Ms. Flora and thus served as a key organization in advancing community-based reuse goals.
3. Most of the city's financial contribution to the project came in the form of an $11,687,766 city bond, which accounted for a third of the total public financing of the project (federal, state, county, and the Pittsburgh Water & Sewer Authority made up the remaining funding).
4. The Waterfront group developed a former steel mill in Homestead, Pennsylvania, just on the edge of the Pittsburgh city limits, with big box retail.
5. In the housing literature, filtering is a process by which landlords invest decreasing amounts of money as their properties age. With this declining quality of housing, lower and lower socio-economic strata occupy these properties as high-income households leave to acquire newer, higher-quality housing (Hoyt 1933; Temkin and Rohe 1996).
6. There were no historic structures on the slag pile at Nine Mile Run.
7. Brownfields such as Washington's Landing and the Pittsburgh Technology Park did not meet the HI-TOADS definition but were both one hundred-plus-acre development that effectively created new communities within the city.
8. The National Parks Service recognizes fifty years as the threshold at which point structures may be considered for their historic value and proposed for listing on the National Register of Historic Places.

7. First Whales, Now Brownfields: New Bedford, Massachusetts (pages 149–75)

1. New Bedford has maintained an active working waterfront despite external efforts to convert the waterfront to a museum, tourist attraction, or residential enclave. Today, the city generates much of its employment and tax revenues from active commercial operations at the waterfront.
2. The task force's decision to combine residential and recreational uses is problematic. Residential uses require that brownfields be cleaned to very high standards, while many recreational uses demand very little remediation due to the lower risks to users.
3. The mayor was credited at the time of the opening of the park with his leadership and support throughout, but city policies and practices played but a small role in the reuse of that HI-TOAD site.
4. The state court system is organized around counties in Massachusetts.
5. See Charles Lindblom's (1959) view of planning as "muddling through."
6. This federal role as a source of innovation in the brownfields arena was documented in a study of the EPA Brownfields Pilot Program (Greenberg and Hollander 2006).

8. Planning for a Shrinking City: Youngstown, Ohio (pages 176–202)

1. Throughout this book, I have used housing data to study neighborhood changes. For those sites that are in primarily commercial districts, it also might be valuable to examine commercial real estate indicators.
2. Some downtown historic structures have been saved.
3. It is worth noting that YSU's new president, David Sweet, was trained in urban studies and public policy and had been the dean of the Levin College of Urban Affairs at Cleveland State University

9. Race, Preservation, and Redevelopment: Richmond, Virginia (pages 203–31)

1. The city continued to use the site for public works storage until 2001.
2. An alternative interpretation is that the influx of new residents had a synergistic effect on existing populations in the neighborhood, which, in turn, elevated the general welfare of all residents. Without more sophisticated analysis, it is not possible to determine precisely how the neighborhood's population changed from 1980 to 2000.
3. In 1999, the property was assessed at $43,000,000 by the City of Richmond's Assessor's Office (Rayner 2001a).
4. NIMBY: Not in my backyard.
5. In New Bedford, for example, four of the five HI-TOADS are owned by the city.
6. The other two cities are Bridgeport, Connecticut, and San Bernardino, California.
7. Based on a report generated on November 11, 2006.
8. This speaks to a limitation of my research design. It is possible that city officials are actually a lot more active behind the scenes in promoting the reuse of HI-TOADS, but are unwilling to admit to it in an interview. I attempted to guard against this by interviewing nongovernmental leaders. In those interviews, I did not learn about any covert effort by city officials to intervene in HI-TOADS.

10. Conclusion (pages 232–51)

1. Richmond was the clear exception to this pattern.
2. These policy recommendations are most useful to cities most similar to those examined in the case studies and during the telephone interviews.
3. Given the growing influence of the "property rights" movement and the Supreme Court's *Kelo* decision, local use of eminent domain may not always be easy.
4. In a world of infinite resources, I would not expect such resistance.

Bibliography

Accordino, J., and G. T. Johnson. 2000. Addressing the vacant and abandoned property problem. *Journal of Urban Affairs* 22, no. 3:301–15.

Andrews, Clint. 1999. The economic and environmental impact of state government on the City of Trenton. Class Report for Industrial Ecology (970:651). Edward J. Bloustein School of Planning & Public Policy, Rutgers, the State University of New Jersey.

Babcock, Richard. 1966. *The zoning game: Municipal practices and policies.* Madison: University of Wisconsin Press.

Bailey, John. 2004. Vacant properties and smart growth: Creating opportunity from abandonment. *Livable Communities @ Work: Funders' Network for Smart Growth and Livable Communities* 1, no. 4.

Bartsch, Charles. 2002. Success story. *Journal of Housing and Community Development* 59, no. 4:18–24.

Bartsch, C., and E. Collaton. 1997. *Brownfields: Cleaning and reusing contaminated properties.* Westport, Conn.: Praeger.

Beauregard, R. A. 2003. *Voices of decline: The postwar fate of U.S. cities.* 2nd edition. New York: Routledge.

Bellah, Robert N. 1985. *Habits of the heart: Individualism and commitment in American life.* Berkeley: University of California Press.

Berke, Philip, and Maria Manta Conroy. 2000. Are we planning for sustainable development? An evaluation of 30 comprehensive plans. *Journal of the American Planning Association* 66, no. 1:21–33.

Bible, Douglas, Chengho Hsieh, Gary Joiner, Chuo-Hsuan Lee, and David Volentine. 2005. Analysis of the effects of contamination by a creosote plant on property values. *The Appraisal Journal* (Winter): 87–97.

Blakely, Edward J. 1994. *Planning local economic development: Theory and practice.* 2nd edition. Thousand Oaks, Calif.: Sage Publications.

———. 2000. Economic development. In *The practice of local government planning,* ed. Charles E. Hoch, 3rd edition. Washington, D.C.: International City/County Management Association.

Bluestone, B., and B. Harrison. 1982. *The deindustrialization of America: Plant closing, community abandonment, and the dismantling of basic industry.* New York: Basic Books.

Board on Infrastructure and the Constructed Environment. 2004. *Investments in federal facilities: Asset management strategies for the 21st century*. Washington, D.C.: The National Academies Press.

Bowman, A. O'M., and M. Pagano. 2004. *Terra incognita: Vacant land and urban strategies*. Washington, D.C.: Georgetown University Press.

Boyer, M. C. 1983. *Dreaming the rational city*. Cambridge, Mass.: The MIT Press.

Brachman, Lavea. 2004. Turning brownfields into community assets: Barriers to re-development. In *Recycling the city: The use and reuse of urban land*, ed. Rosalind Greenstein and Yesim Sungu-Eryilmaz. Cambridge, Mass.: Lincoln Institute of Land Policy.

Bradbury, K. L., A. Downs, and K. A. Small. 1982. *Urban decline and the future of American cities*. Washington, D.C.: Brookings Institution.

Brahler, Steadman. 2005. Could filmmaking become next boom industry? *Business Journal*, August 1.

Bright, E. 2003. Making business a partner in redeveloping abandoned central city property: Is profit a realistic possibility? Federal Reserve System's Third Community Affairs Research Conference. Washington, D.C., March 27–28.

Bright, Elise M. 2000. *Reviving America's forgotten neighborhoods: An investigation of inner city revitalization efforts*. New York: Garland Pub.

Brody, Samuel D., David R. Godschalk, and Raymond J. Burby. 2003. Mandating citizen participation in plan making. *Journal of the American Planning Association* 69, no. 3:245–65.

Burchell, R., and D. Listokin. 1981. *The adaptive reuse handbook*. New Brunswick, N.J.: Center for Urban Policy Research.

Buss, Terry F., and F. Stevens Redburn. 1983. *Shutdown in Youngstown: Public policy for mass unemployment*. Albany: State University of New York Press.

"Buying jobs." 1979. *Time*, December 24.

Byng, Michelle Denise. 1992. A new face in the structure of community power: The black political elite of Richmond, Virginia. Ph.D. diss., University of Virginia.

Byrum, Oliver. 1992. *Old problems in new times: Urban strategies for the 1990s*. Chicago: American Planning Association.

Campbell, Scott, and Susan S. Fainstein. 2001. *Readings in planning theory*. Malden, Mass.: Blackwell Publishing.

Campbell, Scott. 1996. Green cities, growing cities, just cities? Urban planning and the contradictions of sustainable development. *Journal of the American Planning Association* 62 (Summer).

Caro, Robert A. 1974. *The power broker: Robert Moses and the fall of New York*. New York: Vintage Books.

Castells, Manuel. 2000. *The rise of the network society*. Oxford: Blackwell Publishers.

Center for Economic Development. 2002. Regional rankings and Pittsburgh performance. Carnegie Mellon University. March. www.smartpolicy.org/ranks 2002.pdf. Accessed June 14, 2006.

City of New Bedford. 2006. Brownfields—Former Elco Dress Site. Department of Environmental Stewardship. www.ci.new-bedford.ma.us/SERVICES/Environmental /elco.asp. Accessed June 19, 2006.

City of Pittsburgh. 2001. Master development planning in Hazelwood and Junction Hollow. Prepared by the Saratoga Associates for Department of City Planning. December.

———. 2005. Hazelwood Second Avenue design strategy. Prepared by Loysen + Kreuhmeier Architects for Department of City Planning. January.

City of Trenton. 1999. Land use plan. City of Trenton. Adopted by the Planning Board effective January.

City of Youngstown. 2005. Youngstown 2010 citywide plan.

Clark, D. 1989. *Urban decline: The British experience.* London: Routledge.

Cohen, James R. 2001. Abandoned housing: Exploring lessons from Baltimore. *Housing Policy Debate* 12, no. 3:415–48.

Colliers International. 2006. Colliers International US real estate review. www .colliersparrish.com/newsletters/USMarkets.pdf. Accessed May 17, 2006.

Colton, K. W. 2003. *Housing in the twenty-first century: Achieving common ground.* Cambridge, Mass.: Harvard University.

Corey, William. 1996. Proposed New Bedford marine park gains favor. *Standard-Times,* August 2. www.s-t.com/daily/08-96/08-02-96/b0110047.htm. Accessed June 19, 2006.

Corey, William, Jack Stewardson, and Rachel G. Thomas. 1996. Weld fills our stocking: Candidate bears gifts for the area. *Standard-Times.* August 24. www.s-t .com/daily/08-96/08-24-96/a0110004.htm. Accessed June 19, 2006.

Cumbler, John T. 1989. *A social history of economic decline: Business, politics, and work in Trenton.* New Brunswick: Rutgers University Press.

Davis, Scott C. 1988. *The world of Patience Gromes: Making and unmaking a Black community.* Lexington: University Press of Kentucky.

Davis, Todd S., ed. 2002. *Brownfields: A comprehensive guide to redeveloping contaminated property.* 2nd edition. Chicago: American Bar Association.

Dear, M., and S. Flusty. 1998. Postmodern urbanism. *Annals of the Association of American Geographers* 88, no. 1:50–72.

Deason, J. P., G. W. Sherk, and G. A. Carroll. 2001. *Public policies and private decisions affecting the development of brownfields: An analysis of critical factors, relative weights and areal differentials.* Washington, D.C.: U.S. Environmental Protection Agency and the George Washington University.

De Sousa, Christopher A. 2003. Turning brownfields into green space in the City of Toronto. *Landscape and Urban Planning* 62, no. 4:181–98.

———. 2004. Unearthing the benefits of brownfield to green space projects: An examination of perceptions and reactions. Association of Collegiate Schools of Planning 45th Annual Conference. Portland, Oregon, October 21–24.

Dewar, Margaret. 2006. Selling tax-reverted land. *Journal of the American Planning Association* 72, no. 2:167–80.

Dewar, Margaret, and Dabina Deitrick. 2004. The role of community development corporations in brownfield redevelopment. In *Recycling the city: The use and reuse of urban land,* ed. Rosalind Greenstein and Yesim Sungu-Eryilmaz. Cambridge, Mass.: Lincoln Institute of Land Policy.

Dillman, Don A. 1978. *Mail and telephone surveys: the total design method.* New York: Wiley.

Doyle, J. E. 2001. One house at a time. *Journal of Housing and Community Development* 58, no. 1:14–17.

Drape, Joe. 1991. Urban renaissance; An occasional series; Richmond plants hopes on Tobacco Row. *Atlanta Journal and Constitution,* July 7.

Duffy, Robert. 2006. Mayor Duffy's inaugural address. www.ci.rochester.ny.us/index.cfm?id=730. Accessed January 6, 2006.

Economist Intelligence Unit. 2006. Global liveability rankings: 2006. Worldwide Cost of Living Survey. December.

Egan, Timothy. 2005. Ruling sets off tug of war over private property. *New York Times.* July 30, p. 1.

Farber, Stephen. 1998. Undesirable facilities and property values: A summary of empirical studies. *Ecological Economics* 24, no. 1:1–14.

Fitzgerald, Kathleen. 1999. Waiting for the future: Creating new possibilities for Youngstown. Bethesda, Md.: The Harwood Group, for the Charles Stewart Mott Foundation.

Fitzpatrick, Dan. 2000a. The story of urban renewal. *Pittsburgh Post-Gazette.* Business section, May 21. www.post-gazette.com/businessnews/20000521eastliberty1.asp. Accessed June 12, 2006.

———. 2000b. East Liberty then: Initial makeover had dismal results. *Pittsburgh Post-Gazette.* Business section, May 23. www.post-gazette.com/businessnews/20000523intro3.asp. Accessed June 12, 2006.

Friedmann, John. 1987. *Planning in the public domain: From knowledge to action.* Princeton, N.J.: Princeton University Press.

Fuechtmann, Thomas G. 1989. *Steeples and stacks: Religion and steel crisis in Youngstown.* Cambridge: Cambridge University Press.

Gans, H. J. 1962. *The urban villagers: Group and class in the life of Italian-Americans.* New York: Free Press.

Garbarine, Rachelle. 1996. Easing the re-use of once polluted sites. *New York Times,* October 27.

———. 1997. 2d anchor begun in Trenton's revitalization plan. *New York Times,* December 14.

———. 1999. In the region/New Jersey; Hockey arena raises hopes of a livelier Trenton. *New York Times,* November 21.

Gardner, Sarah S. 2001. Green visions for brownfields: The politics of site remediation and redevelopment in four New Jersey cities. Unpublished Ph.D. diss., City University of New York.

Greenberg, Michael R. 1996. *Environmentally devastated neighborhoods: Perceptions, policies, and realities.* New Brunswick, N.J.: Rutgers University Press.

———. 1999. Improving neighborhood quality: A hierarchy of needs. *Housing Policy Debate* 10, no. 3:601–24.

Greenberg, Michael R., Diane Downton, and Henry Mayer. 2003. Are mothballed brownfields sites a major problem? *Public Management* 85, no. 5:12–17.

Greenberg, Michael R., and Justin B. Hollander. 2006. The EPA's brownfields pilot program: A multi-geographically layered and socially desirable innovation. *American Journal of Public Health* 96, no. 2.

Greenberg, Michael R., K. Lowrie, L. Solitare, and L. Duncan. 2000. Brownfields, TOADS, and the struggle for neighborhood redevelopment. *Urban Affairs Review* 35, no. 5:717–34.

Greenberg, Michael R. and F. Popper. 1994. Finding treasure in TOADS. *Planning* 60:24–28.

Greenberg, Michael R., F. Popper, D. Schneider and B. West. 1993. Community organizing to prevent TOADS in the United States. *Community Development Journal* 28:55–65.

Greenberg, Michael R., F. J. Popper, and B. M. West. 1990. The TOADS: A new American urban epidemic. *Urban Affairs Quarterly* 25, no. 3:435–54.

Greenberg, Michael R., F. Popper, B. West, and D. Schneider. 1992. TOADS go to New Jersey: Implications for land use and public health in mid-sized and large US cities. *Urban Studies* 29:117–25.

Greenstein, Rosalind and Yesim Sungu-Eryilmaz. 2004. Introduction: Recycling urban vacant land. In *Recycling the city: The use and reuse of urban land*. Cambridge, Mass.: Lincoln Institute of Land Policy.

Guralnik, David B., ed. 1986. *Webster's new world dictionary of the American language*. New York: Prentice Hall Press.

Hall, Peter. 2000. *Cities of tomorrow: An intellectual history of urban planning and design in the twentieth century*. Malden, Mass.: Blackwell Publishers.

Hanover, Larry. 2006. Broken Hollywood dreams. *The Trenton Times*. February 22.

Harrison, Carolyn, and Gail Davies. 2002. Conserving biodiversity that matters: Practitioners' perspectives on brownfield development and urban nature conservation in London. *Journal of Environmental Management* 65, no. 1:95–108.

Hart, Alison. 2005. Creating and implementing the vision—El Toro case study. NAID/ADC 2005 Annual Conference. Denver, Colorado, June 4–7.

Hartkopf, Sophia. 2005. Southside Works. Unpublished Master's Paper. University of Pittsburgh. December 2.

Hickey, Gordon. 1998. City OKs Shockoe complex. *Richmond Times Dispatch*, May 27.

Hillier, A. E., D. P. Culhane, T. E. Smith, and C. D. Tomlin. 2003. Predicting housing abandonment with the Philadelphia neighborhood information system. *Journal of Urban Affairs* 25, no. 1:91–105.

Hoch, Charles. 1996. A pragmatic inquiry about planning and power. In *Explorations in planning theory*, ed. S. J. Mandelbaum, L. Mazza, and R. W. Burchell. New Brunswick, N.J.: Center for Urban Policy Research.

Hodder, Robert. 1999. Redefining a southern city's heritage: Historic preservation planning, public art, and race in Richmond, Virginia. *Journal of Urban Affairs* 21, no. 4:437–53.

Hoffman, Steven J. 2004. *Race, class and power in the building of Richmond, 1870–1920*. Jefferson, N.C.; London: McFarland & Co.

Hoover, Edgar M., and Raymond Vernon. 1962. *Anatomy of a metropolis: The changing distribution of people and jobs within the New York Metropolitan Region*. Cambridge, Mass.: Harvard University Press.

Hoyt, Homer. 1933. *One hundred years of land values in Chicago*. Chicago: University of Chicago Press.

Humes, Pete. 2003. Where the action is; Tobacco Row lures hipsters and hopesters. *Richmond Times Dispatch,* April 3.

Hurd, Brian H. 2002. Valuing Superfund site cleanup: Evidence of recovering stigmatized property values. *The Appraisal Journal* (October):426–37.

Jackson, Kenneth. 1985. *Crabgrass frontier: The suburbanizaton of the United States.* New York: Oxford University Press.

Kamara, Jemadari. 1983. Planning, plant closings, and public policy. Unpublished Ph.D. diss., University of Michigan, Ann Arbor.

Keating, L., and Sjoquist, D. 2001. Bottom fishing: Emergent policy regarding tax delinquent properties. Fannie Mae Foundation. *Housing Facts and Findings* 3, no. 1.

Keating, W. Dennis, Norman Krumholz, and Philip Star, eds. 1996. *Revitalizing urban neighborhoods.* Lawrence: University Press of Kansas.

Keenan, P., S. Lowe, and S. Spencer. 1999. Housing abandonment in inner cities— The politics of low demand for housing. *Housing Studies* 14, no. 5:703–16.

Kiefer, D. 1980. Housing deterioration, housing codes and rent control. *Urban Studies* 17:53–62

Kiel, Katherine A., and K. T. McClain. 1995. House prices during siting decision stages: The case of an incinerator from rumor through operation. *Journal of Environmental Economics and Management* 28, no. 2:241–55.

Kiel, Katherine A., and Michael Williams. 2006. The impact of Superfund sites on local property values: Are all sites the same? *Journal of Urban Economics* 61: 170–92.

King, Maxwell. 2006. Remarks by Maxwell King, President, The Heinz Endowment. Delaware Valley Grantmakers 18th Annual Meeting. Philadelphia, Pennsylvania. January 11. www.dvg.org/aboutgp/feature_voices.htm. Accessed June 14, 2006.

Kvale, Steinar. 1996. *Interviews: An introduction to qualitative research interviewing.* Thousand Oaks, Calif: Sage Publications.

Lagemann, Ellen C. 1992. *Politics of knowledge: The Carnegie Corporation, philanthropy, and public policy.* Chicago: University of Chicago Press.

Langdon, Philip. 2005a. Eminent domain goes to court. *Planning* 71, no. 4:12–15.

——. 2005b. The not-so-secret code: Across the U.S., form based codes are putting New Urbanist ideas into practice. *Planning* 72, no. 1:24–29.

Lanks, Belinda. 2006. Creative Shrinkage. *New York Times Magazine,* December 10, 40–41.

Leigh, Nancy Green, and Sarah L. Coffin. 2002. Modeling the brownfield relationship to property values and community revitalization. Paper presented at the 5th Symposium of the International Urban Planning and the Environment Association. Oxford, England. September 23–26.

——. 2005. Modeling the brownfield relationship to property values and community revitalization. *Housing Policy Debate* 16:2.

Lindblom, Charles E. 1959. The science of "muddling through." *Public Administration Review* 19, no. 2:79–88.

Lindeman, Teresa F. 2001. A year after opening, Home Depot says its store is doing well, and momentum is beginning to build in the community around it. *Pittsburgh Post-Gazette,* February 4.

Linkon, Sherry Lee. 2002. *Steeltown U.S.A: Work and memory in Youngstown.* Lawrence: University Press of Kansas.

Mallach, Alan. n.d. Putting environmentally contaminated sites back to use: Lessons from Trenton's brownfields experience. Unpublished manuscript.

Marcotte, Bob. 2004. Arsenal of Freedom part one: Rochester products that helped win World War II. *Rochester History* 66, no. 1 (Winter).

Matthews, Anne. 1992. *Where the buffalo roam: Restoring America's Great Plains.* Chicago: University of Chicago Press.

Matthews, William H. 2006. Statement of William H. Matthews, Assistant Commissioner, Office of Real Property Asset Management, Public Buildings Service, U.S. General Services Administration, Before the Committee on Homeland Security and Government Affairs, Subcommittee on Federal Management, Government Information, and International Security, United States Senate. February 6. www.gsa.gov/Portal/gsa/ep/contentView.do?contentType=GSA_Basic& contentID= 20418&noc. Accessed July 11, 2006.

Mayer, Henry, and Michael Greenberg. 2001. Coming back from economic despair: Case studies of small and medium-size American places. *Economic Development Quarterly* 15, no. 3:205–16.

Mayer, Henry, and Judy Shaw. 1997. 467 Calhoun Street, Trenton, New Jersey. Location Project. 970:501. Unpublished class report. Rutgers, The State University of New Jersey.

McCabe, Marsha, and Joseph D. Thomas. 1995. *Not just anywhere: The story of WHALE and the rescue of New Bedford's Waterfront Historic District.* New Bedford, Mass.: Spinner Publications.

McElwaine, Andrew S. 2003. Slag in the park. In *Devastation and renewal: An environmental history of Pittsburgh and its region,* ed. J. Tarr. Pittsburgh: University of Pittsburgh Press.

Meck, Stuart, Paul Wack, and Michelle J. Zimet. 2000. Zoning and subdivision regulations. In *The practice of local government planning,* ed. Charles E. Hoch, 3rd edition. Washington, D.C.: International City/County Management Association.

Melville, Herman. 1851. *Moby-Dick.* New York: Harper & Brothers.

Metzger, J. T. 2000. Planned abandonment: The neighborhood life-cycle theory and national urban policy. *Housing Policy Debate* 11, no. 1:7–40.

Meyer, Peter B., and Thomas S. Lyons. 2000. Lessons from private sector brownfield redevelopers: Planning public support for urban regeneration. *Journal of the American Planning Association* 66, no. 1:46–57.

Miles, Mike E., Gayle Berens, and Marc A. Weiss, eds. 2000. *Real estate development: Principles and process.* 3rd edition. Washington, D.C.: Urban Land Institute.

Miller, Jonathan D. 2005. *Emerging trends in real estate.* Washington, D.C.: Urban Land Institute and PricewaterhouseCoopers LLP.

Miller, Tyler, M. Greenberg, K. Lowrie, H. Mayer, A. Lambiase, R. Novis, D. Ionnides, S. Meideros, and A. Trovato. 2000. Brownfields redevelopment as a tool for smart growth: Analysis of nine New Jersey municipalities. Report 12 for the Office of State Planning. New Brunswick, N.J.: National Center for Neighborhood and Brownfields Redevelopment, Rutgers University.

Mitchell, T. 2002. *Rule of experts: Egypt, techno-politics, modernity.* Berkeley: University of California Press.

Moskowitz, Harvey S., and Carl G. Lindbloom. 1993. *The new illustrated book of development definitions.* New Brunswick, N.J.: Center for Urban Policy Research.

National Oceanic and Atmospheric Administration. 2006. NOAA coastal brownfields. brownfields.noaa.gov/htmls/portfields/portfields.html. Accessed June 19, 2006.

National Park Service. 2006. New Bedford whaling. www.nps.gov/nebe. Accessed June 26, 2006.

National Research Council. 1994. *Ranking hazardous-waste sites for remedial action. Committee on Remedial Action Priorities for Hazardous Waste Sites.* Board on Environmental Studies and Toxicology, Commission on Geosciences, Environment, and Resources. Washington, DC: National Academies Press.

New Bedford Chamber of Commerce. 1997. Report of the New Bedford Chamber of Commerce Brownfields Working Group. September.

NJDEP. 2006a. Brownfields Development Area (BDA) initiative. New Jersey Department of Environmental Protection. www.nj.gov/dep/srp/brownfields/bda/. Accessed January 4, 2006.

———. 2006b. New Jersey brownfield projects receive national Phoenix Awards showcasing excellence in cleanup and redevelopment. www.state.nj.us/dep/srp/publications/brownfields. Accessed May 9, 2006.

Nicodemus, Aaron. 2003. City assures that Pierce Mill will be a park. *Standard-Times,* 11 August. www.s-t.com/daily/08-03/08-11-03/a0910052.htm. Accessed June 19, 2006.

———. 2005. Finally, a walk in the park: Decade of effort pays off as Riverside Park opens. *Standard-Times.* 3 November. www.s-t.com/daily/11-05/11-03-05/ a0110726 .htm. Accessed June 19, 2006.

Northam, Ray. 1971. Vacant urban land in the American city. *Land Economics* 47: 345–55.

Odland, J., and B. Balzer. 1978. Localized externalities, contagious processes and the deterioration of urban housing: An empirical analysis. *Socio-Economic Planning Sciences* 13:87–93.

O'Toole, Christine H. 2005. Arts and science remake the steel city. *New York Times,* Section C, July 19.

Pagano, M. A. and A. O. Bowman. 2000. Vacant land in cities: An urban resource. Survey series for the Brookings Institution Center on Urban and Metropolitan Policy and CEOs for Cities. December.

Page, G. William, and Harvey Rabinowitz. 1993. Groundwater contamination: Its effects on property values and cities. *Journal of the American Planning Association* 59:473–81.

Parker, L. A. 2004. Developer presents $13M housing plan. *The Trentonian,* February 4.

Pattison, G. 2004. Planning for decline: The "D"-village policy of County Durham, UK. *Planning Perspectives* 19 (July):311–32.

Pearce, David, Anil Markandya, and Edward B. Barbier. 1989. *Blueprint for a green economy: A report.* London: Earthscan.

Platt, Kevin. 2000. Going green. *Planning* 66, no. 8:18–22.

Platt, Rutherford H. 1996. *Land use and society: Geography, law and public policy.* Washington, D.C.: Island Press.

Popper, D. E., and F. J. Popper. 2002. Small can be beautiful: Coming to terms with decline. *Planning* 68, no. 7:20–23.

Popper, Frank. 1981. Siting LULUs. *Planning* 57.

Portney, Kent. 2002. Taking sustainable cities seriously: A comparative analysis of twenty-four U.S. cities. *Local Environment* 7, no. 4:363–80.

Powers, C., F. Hoffman, D. Brown, and C. Conner. 2000. *Experiment: Brownfields pilots catalyze revitalization.* New Brunswick, N.J.: Institute for Responsible Management.

Rayner, Bob. 2001a. Hospital site finds buyer; Lutherans plan retirement facility for North Side. *Richmond Times Dispatch,* January 23.

———. 2001b. Plan for former hospital site falls through. *Richmond Times Dispatch,* November 28.

———. 2001c. For sale again: Hospital back on the block after plan for retirement community fails. *Richmond Times Dispatch,* December 10.

———. 2002. Return of the native: Developer's firm to renovate hospital where he was born. *Richmond Times Dispatch,* March 1.

Real Estate Research Corporation. 1975. The Dynamics of Neighborhood Change. Washington, D.C.: U.S. Department of Housing and Urban Development, Office of Policy Development and Research.

Reichert, Alan K., Michael Small, and Sunil Mohanty. 1992. The impact of landfills on residential property values. *Journal of Real Estate Research* 7, no. 3:297–314.

Ress, David. 2004. Council to consider plan to develop gas works site: Fulton deal would let city, developers share the risks and profits. *Richmond Times Dispatch,* January 26.

Rodwin, Lloyd, and Bishwapriya Sanyal, eds. 2000. *The profession of city planning: Changes, images and challenges 1950–2000.* New Brunswick, N.J.: Center for Urban Policy Research.

Rohe, William M., and Leslie S. Stewart. 1996. Homeownership and neighborhood stability. *Housing Policy Debate* 7, no. 1:37–81.

Rusk, D. 1995. *Cities without suburbs.* 2nd edition. Washington, D.C.: Woodrow Wilson Center Press.

Safford, Sean C. 2004. Why the garden club couldn't save Youngstown: Social embeddedness and the transformation of the Rust Belt. Unpublished Ph.D. diss., MIT, Cambridge, Mass.

Salzman, Avi, and Laura Mansnerus. 2005. For homeowners, frustration and anger at court ruling. *New York Times,* June 24, 20.

Sassen, Saskia. 1991. *The global city: New York, London, Tokyo.* Princeton, N.J.: Princeton University Press.

Savageau, David. 2007. *Places rated almanac.* 7th ed. Washington, D.C.: Places Rated Books, LLC.

Schattle, Hans. 1998. Challenges of public deliberation: A case study of citizen participation in environmental policy. Master's thesis, Boston College.

Schawb, K., and P. Middendorff. 2003. Revitalising the urban core. *Economic Development Journal* 2, no. 2:24–30.

Schneider, Paul. 2006. New Bedford, Mass. *New York Times,* May 26, 3. Accessed via LexisNexis Academic on June 23, 2006.

Self, Robert O. 2003. *American Babylon: Race and the struggle for postwar Oakland.* Princeton, N.J.: Princeton University Press.

Shea, Kevin. 2006. Man found dead was city resident. *Trenton Times,* March 23, A4.

Silver, Christopher. 1983. The ordeal of city planning in postwar Richmond, Virginia: A quest for greatness. *Journal of Urban History* 10, no. 1:33–60.

———. 1984. *Twentieth-century Richmond: Planning, politics, and race.* Twentieth-Century America Series. Knoxville: University of Tennessee Press.

Silver, Christopher, and John V. Moeser. 1995. *The separate city: Black communities in the urban South, 1940–1968.* Lexington: University Press of Kentucky.

Simons, Robert. 1999. How many brownfields are out there? An economic base contraction analysis of 31 U.S. cities. *Public Works Management and Policy* 2, no. 3:267–73.

Simons, Robert A. 2006. *When bad things happen to good property.* Washington, D.C.: Environmental Law Institute.

Simons, Robert A., and Jesse Saginor. 2006a. A meta-analysis of the effects of environmental contamination on residential real estate values. In *When bad things happen to good property,* ed. Robert A. Simons. Washington, D.C.: Environmental Law Institute.

———. 2006b. A meta-analysis of the effects of environmental contamination and positive amenities on residential real estate values. *Journal of Real Estate Research* 28, no. 1:71–104.

Sites, William. 1997. The limits of urban regime theory: New York City under Koch, Dinkins, and Giuliani. *Urban Affairs Review* 32, no. 4:536–57.

Smith, N. 1996. *The new urban frontier: Gentrification and the revanchist city.* New York: Routledge.

Smith, Roger G. 2002. City seeks studies of long-vacant properties. *The Vindicator,* December 26.

Stikkers, David, and Joel A. Tarr. 1999. Pittsburgh's brownfields: Case studies in redevelopment. Unpublished manuscript.

Stipe, R. E., ed. 2003. *A richer heritage: Historic preservation in the twenty-first century.* Chapel Hill: University of North Carolina Press.

Stone, C. N. 1989. *Regime politics: The governing of Atlanta 1946–1988.* Lawrence: University Press of Kansas.

Strauss, A., and J. Corbin. 1998. *Basics of qualitative research: Techniques and procedures for developing grounded theory.* 2nd edition. Thousand Oaks, Calif.: Sage Publications.

Tarr, Joel. 2002. The metabolism of the industrial city: The case of Pittsburgh. *Journal of Urban History* 28, no. 5:511–45.

———. 2003. Introduction: Some thoughts about the Pittsburgh environment. In *Devastation and renewal: An environmental history of Pittsburgh and its region.* Pittsburgh: University of Pittsburgh Press.

Teaford, J. C. 2000. Urban renewal and its aftermath. *Housing Policy Debate* 11, no. 2:443–65.

Temkin, Kenneth, and William Rohe. 1996. Neighborhood change and urban policy. *Journal of Planning Education and Research* 15:159–70.

Thomas, June Manning, and J. Eugene Grigsby III. 2000. Community development. In *The Practice of Local Government Planning*, ed. Charles E. Hoch, 3rd edition. Washington, D.C.: International City/County Management Association.

Tregoning, Harriet, Julian Agyeman, and Christine Shenot. 2002. Sprawl, smart growth and sustainability. *Local Environment* 7, no. 4:341–47.

Tyler-McGraw, M. 1994. *At the falls: Richmond, Virginia, and its people.* Chapel Hill: University of North Carolina Press.

U.N. World Commission on Environment and Development. 1987. *Our common future.* Oxford: Oxford University Press.

U.S. Census. 2000. American FactFinder. Decennial Census. www.census.gov. Accessed December 10, 2004.

U.S. Department of Housing and Urban Development, Office of Policy Development and Research. 2004. U.S. Housing Market Conditions, May.

U.S. Environmental Protection Agency. 1998. Mustard plants helping to clean-up site. Brownfields Success Stories. Outreach and Special Projects Staff. EPA-500-F-98-xxx. http://www.epa.gov/brownfields/pdf/ss_trnt1.pdf. Accessed April 17, 2006.

U.S. Environmental Protection Agency. 2000. Brownfields data primer.

U.S. Environmental Protection Agency. 2003. Local economic impacts of site redevelopment. Washington, D.C.

U.S. Environmental Protection Agency. 2006a. Crane properties fact sheets. Region 2 Superfund. www.epa.gov/cgi-bin/epaprintonly.cgi. Accessed April 17, 2006.

U.S. Environmental Protection Agency. 2006b. Brownfields success stories: From slag to riches. www.epa.gov/swerosps/bf/html-doc/ss_pitts.htm. Accessed June 15, 2006.

U.S. Federal Home Loan Bank Board. 1940. Waverly: A study in neighborhood conservation.

U.S. General Accounting Office. 1978. Report to Congress of the United States by the comptroller general: Housing abandonment: A national problem needing new approaches. August 10.

van Vliet, Willem, ed. 1997. *Affordable housing and urban redevelopment in the United States.* Thousand Oaks, Calif.: Sage Publications.

Vergara, C. J. 1995. *The new American ghetto.* New Brunswick, N.J.: Rutgers University Press.

Vey, Jennifer S. 2006. Revitalizing weak market cities in the U.S. 16th Annual Conference on the Small City. Wilkes-Barre, Pennsylvania.

Voyer, Richard A., Carol Pesch, Jonathan Garber, Jane Copeland, and Randy Comeleo. 2000. New Bedford, Massachusetts: A story of urbanization and ecological connections. *Environmental History* 5, no. 3:352–77.

Wallace, R. 1989. Homelessness, contagious destruction of housing and municipal service cuts in New York City: 1. Demographics of a housing deficit. *Environment and Planning* 21:1285–1603.

Ward, Kevin. 1996. Rereading urban regime theory: A sympathetic critique. *Geoforum* 27, no. 4:427–38.

Wernstedt, Kris. 2004. Overview of existing studies on community impacts of land reuse. Working Paper 04–06, National Center for Environmental Economics, U.S. Environmental Protection Agency, Washington, D.C.

Wheatley, Paul. 2000. The passage of state Issue 1 provides money for brownfields revitalization. *The Vindicator,* November 19.

White, C. Langdon, Edwin J. Foscue, and Tom L. McKnight. 1964. *Regional geography of Anglo-America.* Englewood Cliffs, N.J.: Prentice-Hall.

Williams, Michael Paul. 2003. Downtown's successes fuel feeling of optimism. *Richmond Times Dispatch,* June 11.

Wilson, D., and H. Margulis. 1994. Spatial aspects of housing abandonment in the 1990s: The Cleveland experience. *Housing Studies* 9, no. 4:493–511.

Wilson, J. Q., and G. L. Kelling. 1982. Broken windows. *Atlantic Monthly* 249, no. 3: 29–38.

Wolfbein, Seymour L. 1944. *The decline of a cotton textile city: A study of New Bedford.* New York: Columbia University Press.

Yin, Robert K. 1994. *Case study research: Design and methods.* Thousand Oaks, Calif.: Sage Publications.

Young, I. M. 2000. *Inclusion and democracy.* Oxford: Oxford University Press.

Yount, Kristen R., and Peter B. Meyer. 1994. Bankers, developers, and new investment in brownfield sites: Environmental concerns and the social psychology of risk. *Economic Development Quarterly* 8, no. 4:338–44.

YSU. 2001. Making something new out of the abandoned steel mill land (brownfields) along the Mahoning River Corridor. Youngstown State University Public Service Institute. September.

Zelinka, Al, and Jennifer Gates. 2005. Money matters. *Planning* 71, no. 2:28–31.

Index

abandoned buildings, 3, 4, 5
ACTION, 193
ALMONO, 133, 140, 147
American Institute of Certified Planners, 21, 122
American Planning Association, 21
arson, 45
asset-based planning, 59–60, 236
Association of Collegiate Schools of Planning, 21

Black Monday, 176–77, 200
bottom-feeder, 41, 174
Brookings Institution, 222
Brownfield Opportunity Area, 55
Brownfields Development Area (BDA), 53–54, 110
Brownfields Environmental Solutions for Trenton (BEST), 84, 92, 107, 112
Brownfields Pilot Program, 27, 60, 106, 139, 296
brownlining, 44, 294
Bullard Street Neighborhood Association, 163

CERCLA, 56, 293
Clean Ohio, 192
Community Reinvestment Act, 24
community-based organizations (CBOs), 24, 76
crime, 38, 44

deindustrialization, 14
Delco Appliance Factory, 1
development review, 53

Early Warning System (EWS), 245
East Liberty Quarter Chamber of Commerce, 134–35, 146
eminent domain, 23, 51, 52
Empire Zones, 53–55
Environmental Cleanup Responsibility Act, 41, 83
Euclidean zoning model, 50

factor analysis, 31
Federal Emergency Management Agency (FEMA), 99, 100
form-based planning/zoning, 50, 74

Geolytics, 249
Greater Fulton Civic Association, 214, 226

Hands Across the River, 167
Hazard Ranking System (HRS), 11, 247
Hazardous Discharge Site Remediation Fund (HDSRF), 83
historic preservation, 74
Human Subjects Review, 294

illegal dumping, 45
Industrial Reuse Ordinance Floating Districts (I-RODS), 50
Industrial Site Recovery Act (ISRA), 83
Institutional Review Board, 294
Isles, Inc., 87, 92, 107, 114, 115

Kelo v. City of New London, 17, 51, 52, 297

Landfills, 37
Locally Unwanted Land Uses (LULUs), 3, 4, 42, 60

Mahoning River Corridor of Opportunity
    (MRCO), 188–90, 194–96
Mallach, Alan, 83
Manex, 98
Mayor's Neighborhood Roundtable, 224
Mercer County Improvement Authority, 105
mothballing, 26
Murphy, Tom, 120, 138

National Brownfields Association, 165
National Register of Historic Places, 222,
    223, 296
neighborhood life-cycle theory, 16–17, 23,
    54, 73, 233, 244
New Bedford Chamber of Commerce, 165
New Jersey Department of Environmental
    Protection (NJDEP), 87, 105, 114
New Jersey Housing Mortgage Finance
    Agency, 97
New Urbanists, 50
non-governmental organizations (NGOs), 7
Northwest Community Improvement
    Association, 87, 107

Oakland, California, 13
Old Bedford Village, 160, 167, 173

Palmer, Doug, 83
Pawtucket, Rhode Island, 28
Planned Development Districts (PDDs), 49
Portfields, 167
Portland, Oregon, 28
power plants, 37

Salvation Army, 65
Scranton, 75
shovel-ready, 48, 55–57, 237
slotting process, 165, 170
smart decline, 202, 252
snowball-sampling technique, 33
South Side Local Development Company,
    122, 126, 139, 146
Spearman's rank correlation, 35
Superfund, 8, 166, 180

tax delinquency, 23
TOADS, 2, 5
Trend MLS, 91
Trenton Roebling Community
    Development Corporation (TRCDC),
    97, 98, 107

Urban Redevelopment Authority (URA),
    119, 122, 126–29, 131, 134–36, 142
urban regime theory, 22

vacant lots, 3–6

Waterfront Historic Area League (WHALE),
    151, 168
weak-market city, 222
white flight, 83

Youngstown 2010 Citywide Plan, 176, 194,
    195, 202

zoning, 49, 69